D1453091

WATER

Edible

Series Editor: Andrew F. Smith

EDIBLE is a revolutionary series of books dedicated to food and drink that explores the rich history of cuisine. Each book reveals the global history and culture of one type of food or beverage.

Already published

Water

A Global History

Ian Miller

REAKTION BOOKS

TD
345
.M635
2015

Published by Reaktion Books Ltd
33 Great Sutton Street
London EC1V 0DX, UK
www.reaktionbooks.co.uk

H

First published 2015

Printed and bound in China by Toppan Printing Co. Ltd

A catalogue record for this book is available
from the British Library

ISBN 978 1 78023 501 1

I

What is Water?

Water seems to be diffused everywhere, and to
be present in all space wherever there is matter.
There are hardly any bodies in nature but what will
yield Water: it is even asserted that fire itself is not
without it. A single grain of the fiery salt, which in
a moment's time will penetrate through a man's hand,
readily imbibes half its weight of Water, and melts even
in the driest air imaginable.

Dr Hutton's Mathematical and Philosophical Dictionary (1796)

Water is ubiquitous. Not only does it cover approximately 70 per cent of the world's surface, but it is considered essential for the survival of all known life forms. The human body is replete with the substance, with a water content as high as 70 per cent. Water aids human digestion, absorbs nutrient, assists blood circulation, removes toxins and regulates temperature. It also transports oxygen to cells and helps to protect joints and organs. Given water's global omnipresence and its vital role in sustaining health, it might be tempting to presume that acquiring access to it for drinking purposes is relatively straightforward. Yet in reality, human societies have faced perpetual difficulties in obtaining consumable water.

Poster of balanced meals featuring 'water: the only indispensable drink', Department of Sanitation and Social Affairs, Martinique.

Throughout time humans have invented an array of drinks designed to offer alternatives to water. However, none of these have ever truly proved to be suitable substitutes. Despite the pervasive popularity of beverages including tea,

coffee and alcohol – liquids denigrated as 'artificial drinks' by nineteenth-century health promoters – the healthy digestion of these products has always depended upon the body containing adequate levels of water. Many man-made alcoholic and soft drinks have dehydrating effects.

Nonetheless, despite widespread public awareness of the substance's physiological importance, water fails to appeal in quite the same way as these more appetising, tasty drinks. Water has no distinctive smell that draws us to it. Nor is its dull, unexciting and transparent appearance particularly interesting or appealing. The French writer and aviator Antoine de Saint-Exupéry recognized this in 1939 when he wrote:

> Water, thou hast no taste, no color, no odor; canst not be defined, art relished while ever mysterious. Not necessary to life, but rather life itself, thou fillest us with a gratification that exceeds the delight of the senses.

In essence water is a tasteless mixture of oxygen and hydrogen atoms that are chemically bonded together. Adjust its temperature and it transforms into ice, steam or vapour. When viewed in this somewhat clinical fashion, water hardly presents itself as an appetizing consumable substance. It does little to lift our mood in the way that caffeine-infused drinks do. Nor does it intoxicate us as alcohol does. Instead, thirst and necessity have mostly dictated human decisions to drink water. But water is something more than this strange mixture of elements and has always done something mysteriously fulfilling to our bodies and minds, as Saint-Exupéry recognized. The twentieth-century English novelist D. H. Lawrence also appreciated this when he observed that 'water is H_2O, hydrogen two parts, oxygen one, but there is also a third thing, that makes water and nobody knows what that is.' Lawrence

perceptively understood that despite its seemingly bland nature, water serves a mysterious but important purpose in human existence. Nonetheless its tasteless, odourless appearance makes it something of a curiosity among drinks and profoundly differentiates it from other potable liquids.

From around the seventeenth century health promoters have laboriously striven to impress upon the public a sense of the necessity of regular water consumption. They have done so because of their conviction that we need to drink approximately half the amount of water in our bodies by weight each day. Should these quantities fail to be ingested, so we are frequently warned, chronic cellular dehydration may result, a complaint that weakens immune systems and causes chemical and nutritional imbalance. In response to concerns such as these, we are nowadays bombarded with messages informing us about the particular amounts of water that we should consume and how our bodies will inevitably suffer should we fail to do so.

But imagine a world where water did not flow freely from taps, or where it could not be readily purchased from a local supermarket or shop. This scenario might seem alien. Yet this was a world familiar to all just 150 years or so ago, even in the most industrialized and modern countries. At that time gaining access to drinking water was only part of the problem. The safety of water, even if it appeared clean, could rarely be depended upon. Today we express disgust and anxiety if we suspect that pollution, pesticides or fertilizers have somehow infiltrated our water supplies. Our predecessors had other concerns. Instead they fretted about whether the water available to them for drinking purposes carried the germs of life-threatening diseases such as cholera and typhoid, or if it contained living animals or remnants of raw sewage.

'Drink water when you are thirsty', information poster, Norway.

Satellite image
of the River Nile
taken at Cairo,
Egypt.

The nineteenth century was a world without germs, in
the sense that science was yet to discover them. Even when
scientists established their potential presence in communal
water supplies in the late nineteenth century, their controver-
sial suggestion that the accidental consumption of certain
bacteria was directly related to the onset of diseases, including cholera, was accepted only gradually. In contrast, we now
typically consider the suitability of drinking water in terms of
its purity and freedom from pollution and germs. However,
it is only throughout the last hundred years or so that precise
understandings of the often disturbing microbial and bacterial content of water have been more fully elucidated. Water
has become consumable on a large scale only relatively recently
and only in consequence of human technological intervention.
Rather than letting the unsettling microbacterial discoveries
made during the nineteenth century deter us from ingesting
water, governments, scientists and businesses instead set

measures in place to guarantee widespread public access to safe consumable supplies.

This outcome has been predominantly restricted to Western societies. Bangladesh, for instance, still suffers from severe rural and urban water scarcity, as well as problems in ensuring the availability of safe drinking water. There, diarrhoeal diseases kill over 100,000 children each year. In 1993 scientists discovered that the extensive well system constructed across Bangladesh in the 1970s had in fact long been contaminated with dangerously high levels of arsenic. The long-term problems associated with gradual arsenic intake that surfaced included skin disorders and internal cancers. In rural Ethiopia, meanwhile, women and children have for some centuries walked for up to six hours to collect water from unprotected ponds, which they share with animals. The country has experienced recurrent droughts, which have caused alarmingly high mortality levels from water-related diseases. Similarly around half of Haiti's population still lack access to clean water. Most Indian water sources are contaminated by sewage and agricultural runoff, the upshot being that around 21 per cent of communicable diseases in India can be traced to the consumption of unsafe water. Western non-governmental organizations including Water.org have taken considerable steps to ameliorate these conditions globally. To achieve this they have referred to the extensive body of knowledge relating to water that has been amassed by medical and scientific experts over the last few centuries.

But is the water consumed daily in Western societies really as 'pure' as it is thought to be? What is actually in our drinking water? Since the industrial age, Western scientists have successfully developed initiatives to ensure its purity and cleanliness by instilling toxins, solids, disease vectors and pathogens into supplies, additions that serve to protect humans from

Information poster advising the boiling of drinking water, Health Education Division, Upjohn, Kenya.

Poster advising drinking boiled water with sugar and salt added to treat diarrhoea, New Caledonia, Commission du Pacifique Sud, 1987.

inadvertently swallowing harmful, unpleasant substances and germs. Today we consider our water supplies in terms of cleanliness and purity and, in many instances, perceive tapped water as a natural substance helpfully piped into our homes. But ironically our modern-day obsession with securing 'pure' drinking water obscures the amount of additional substances that have been infused into our supplies to render it consumable. In many ways, tapped and even bottled water are entirely different substances from 'natural' water.

The irony of this scenario was not lost on early twentieth-century anti-chemical campaigners. In 1955 the prominent American alternative health promoter Raymond W. Bernard provocatively asked:

> When you turn on your kitchen faucet and observe the innocent-looking transparent fluid emerge from it, have you ever stopped to think that this water may really be a wolf in sheep's clothing, and that rather than being pure water, it is really a dilute solution of chemicals; which, while it may not sicken you or kill you outright, might, however, have a long-standing cumulative effect that may lead to eventual disease?

Bernard continued by asserting ominously that

> Water, which is man's best friend, has today become his worst enemy. It is full of chemicals added to it in order to 'purify' it, as well as metals dissolved from the pipes through which it passes. When you put this water in a pot and cook it, you boil it down and so increase the concentration of the chemicals and metals it contains – so that rather than purify water, the boiling of city water makes it more poisonous than it otherwise is.

Water's invisible content has always been feared. In recent times this anxiety has been expressed through public anxieties about the detrimental bodily effects of ingesting germs and purifying chemicals. Curiously, however, water has always been simultaneously glorified and venerated despite a perpetual apprehension about the substance. According to the fifth-century BC Greek philosopher Pindar, 'the best thing is water, and the next gold.' For the pioneering eighteenth-century Scottish physician George Cheyne, 'without all, peradventure, water was the primitive – the original beverage, and it is the only fluid fitted for the ends appointed by nature. Happy had it been for the race of mankind if other mixed and artificial liquors had never been invented.'

Humans are obsessed with water. It has always fascinated us. Today we fanatically strive to ensure the purity of our supplies, and feel deep concern, as Bernard did, when we suspect that its ostensibly pure, transparent appearance may

A row of drinking-water vending machines in Pattaya, Thailand.

be concealing harmful hidden contents likely to injure our health if swallowed. Water is healthy, natural and necessary for our bodies. At the same time, it has the potential to be poisonous, polluted and contaminated. This has ensured that, historically, ideas about drinking water have tended to be multilayered. They have simultaneously privileged the importance of drinking water while warning of its potential dangers.

What, then, constitutes good water? When the prominent American chemist Charles Frederick Chandler delivered his lecture on water to the American Institute of the City of New York in 1871, he informed his audience that

> The characteristics of a good drinking water may be enumerated as follows: The temperature should be at least ten degrees lower than the temperature of the atmosphere, but it should not be much lower than forty-five degrees Fahrenheit. It should be free from taste, except, perhaps, a slight pungency from oxygen and carbonic acid, which is an advantage. Taste is, however, a poor guide. When one becomes accustomed to a certain water, pure water tastes flat by comparison; fifty grains of chloride of sodium in a gallon would hardly affect the taste perceptibly. A third requirement is freedom from smell. This should not be apparent, even when a bottle is half filled with the water, placed in a warm place for a few hours, and then shaken. It should be transparent; not that it is necessarily poisonous if not transparent, but it is preferable to take our solid food in other forms. Sometimes water may contain peaty matter from swamps, or vegetable matter from new reservoirs which is not necessarily unwholesome.

Evidently Chandler, although attentive to important considerations including taste and smell, perceptively recognized

that water may contain mysterious additions that are invisible to the naked human eye. In his view water often held a hidden realm of potentially harmful substances, even when it appeared to be fresh. It may have been polluted by natural matter that did not necessarily kill or cause illness if ingested, but which hardly seemed good for the human body. And even the types of water that humans naturally developed a taste for were not always the purest, according to Chandler. Clearly to probe into the history of drinking water requires inquiring into a history of confusing, contradictory messages that have constantly shifted over time.

2

Water Falls in and out of Fashion

Ancient civilizations gathered around water sources. They did so because they instinctively recognized water's necessity to human existence while still appreciating that not all water was consumable. When settling they sought out sites that were close to the cleanest, healthiest and most refreshing water supplies. Proof of this was afforded in the eighteenth century when, after sipping fresh water from the River Nile (along which the ancient Egyptian civilization had developed), the French Consul General Benoît de Maillet felt so inspired that upon returning home he penned a short essay entitled 'On the Extraordinary and Extreme Deliciousness of the Waters of the Nile'. In his essay Malliet enthusiastically declared:

> When one drinks of it the first time, it seems to be some water prepared by art. It has something in it inexpressibly agreeable and pleasing to the taste; and we ought to give it perhaps the same rank among waters which Champagne has among wines.

Ancient societies congregated around water sources such as the Nile. And if water supplies were not quite as close as might be hoped for, then, in some instances, they were

simply moved. Incan communities developed the capacity to shift water from far distant springs into their impressive capital Machu Picchu, located at a staggering height of over 7,000 feet. They achieved this by constructing a complex and extensive network of sloping canals, fountains and agricultural terraces. Other ancient communities developed an array of technological and social practices designed to ensure the availability of clean water. Ancient Egyptians carved sketches into their tombs which have left evidence of their technological capacity to siphon out water impurities. In 200 BC the ancient Mesopotamians implemented public sanitation laws designed to segregate cisterns and wells from potential sources of contamination, including cemeteries, tanneries and slaughterhouses.

It is the ancient Romans, however, who continue to impress us even today with their strikingly intricate technological systems of water management. Recognizing the importance of

Yakhcal of Yazd province, an ancient type of evaporative cooler.

Egyptian women with jars carrying water from the Nile, 1900–1920.

clean water, Roman emperors famously set their slaves to work constructing an impressive aqueduct system that supplied major cities across their expansive empire with consumable water. Nine different aqueducts delivered water to the city of Rome alone, while surrounding reservoirs, pipes and public fountains enabled an efficient system of storage and distribution. It is little wonder that the late Victorians, when laying the foundations of a modern water supply system in Britain, routinely referred back to, marvelled at and strove to emulate Roman achievements.

One impressed author, writing in 1870 at the peak of Britain's Industrial Revolution in the popular literary and political *London Quarterly Review*, compared nineteenth-century water engineering endeavours to those undertaken by the Romans, asserting that

In some of the most important improvements of the present day – in the supply of good drinking water and

in the construction of public baths – we are now only going over the same ground as Ancient Rome. That city and indeed all the Roman colonies were well supplied with water, often brought from a distance at a vast expense; and the remains of the public baths in Rome and in large provincial cities, of those attached to private villas in Rome and even in its more remote settlements, are on a scale quite beyond anything attempted in modern times.

Nonetheless, despite these technological achievements, alcohol retained its position as a universally popular beverage across the expansive Roman Empire. In fact, as further chapters will demonstrate, alcohol and water were to share an intimate and interconnected history for many centuries to come. Of course the ancient popularity of beverages such as wine can be partly ascribed to their intoxicating alcoholic content. But water could also be bought and vended as a

The ancient Roman Pont du Gard aqueduct in southern France, built AD 40–60.

drinkable commodity, as was the case in the city of Rome. Although public basins were widely available to residents of that city, special water taxes were imposed on those who attached water pipes to their home for the purpose of domestic consumption. To have running drinking water at home was then considered a luxury, and remained so until relatively recently.

Evidently water played a pivotal role in the day-to-day life of the expansive empires that emerged and evolved in the ancient period. The need to have access to consumable supplies played a powerful part in shaping the geographical evolution of political territories, their social functioning and the physical experience of being incorporated into an empire. In many ways these considerations still bear relevance. The Russian president Mikhail Gorbachev, leader of a very different but still expansive country, declared towards the close of the twentieth century that

> Water, like religion and ideology, has the power to move millions of people. Since the very birth of human civilization, people have moved to settle close to it. People move when there is too little of it. People move when there is too much of it. People journey down it. People write, sing and dance about it. People fight over it. And all people everywhere and every day need it.

Given this, it is difficult to contemplate how the practice of using water for drinking purposes fell out of fashion as the Roman Empire declined. Enemies of the empire knew that Roman society could be swiftly reduced by attacking its key social amenities. Accordingly, when Goth besiegers attacked Rome, they attacked and smashed down its extensive aqueduct system, reducing medieval Romans to drawing polluted

water from wells or alternatively from the River Tiber, a water source then also used for sewerage purposes.

For some centuries Roman techniques of transporting clean water remained mostly forgotten. When Rome's empire fell, few individuals cared to, or knew how to, maintain its striking aqueduct system; one corollary was a retreat to traditional but often less reliable sources. It was not uncommon for residents of medieval towns and cities to exploit the same water resources for drinking and sewerage purposes. This hardly bolstered water's reputation as a consumable substance. Occasional glimmers of hope surfaced in the form of urban water distribution schemes. In 1236 Henry II granted permission for the laying of elm and lead pipes to carry water beneath the streets of London and directly into the homes of the wealthy. Yet those who could not afford this privilege – the vast majority of London's population – continued to depend upon unsanitary water drawn from the River Thames, although some merchants and water companies even attempted to charge for the use of that.

Five hundred years later, this unfortunate situation hardly seemed to have improved. The eighteenth-century German travel writer Johann Georg Keyssler left Paris disgusted after observing how

> The two pumps of La Samaritaine and Pont de Nôtre Dame, supply the city with a great quantity of water; but it is only from the river and after it has run through half the city, and thereby becomes very foul. The remote parts labour under the inconveniency of purchasing this water from the *porteurs d'eau*, or water carriers.

At worst water came to be associated in the public eye with filth, disease and death. It is in this context that ales, meads

and wines acquired considerable popularity. In addition to being intoxicating, alcoholic drinks were processed. Moreover, alcohol kills germs, rendering their consumption relatively safe. Water, on the other hand, was often polluted, infected and highly dangerous.

Evidently a sense that citizens had the right to access sanitary water had yet to come into existence. Our medieval counterparts would not have been particularly convinced by modern conceptions of water as a bodily necessity and health-sustaining product. It was not that everyone abstained completely from drinking water, but cautious attitudes undeniably existed towards the practice. Given the choice, many individuals chose to drink more reliable consumables. In many ways the remainder of this book reconstructs past efforts to make water safe, to rekindle interest in it as a vital liquid consumable and to convince the public of water's physiological utility and necessity.

3
What is in Water?

Can water be transformed into a substance that is safe, as well as pleasant, to consume? Early modern scientists and philosophers pondered this question intently. At that time, collecting rainwater for use as drinking water seemed out of the question. As the English physician Thomas Percival warned in his *Experiments and Observations on Water* of 1769:

> Rain water, when collected in clean vessels, is light, soft and wholesome. But as it passes through the atmosphere, which is a chaos of different exhalations from the animal, vegetable and mineral kingdoms; it must wash down some of those floating, volatile particles, and be impregnated with them.

Snow water, if you could find it, was deemed slightly more wholesome, although scientists warned that it often contained minute traces of nitrous acid. Spring waters were also looked upon favourably, although they were known occasionally to contain animal and vegetable mucilage. Scientists deemed seawater to be an entirely unsuitable drink. Because of its high salt content, seawater intensifies thirst, although some early modern scientists pinpointed an array of other problems with

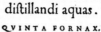

The distillation of water in Pietro Andrea Matthioli's *Senesis medici*, 16th century.

it. In the 1770s the Swedish chemist Torbern Bergman declared that the only types of seawater that tasted pleasant were those found at a depth of at least 60 fathoms since, so Bergman believed, water collected from above that level was crammed full of the disintegrating remains of putrefying fish and sea creatures which had risen to the surface and caused contamination.

Evidently water was historically recognized as a liquid substance with varying properties and content. Water's suitability for consumption was typically assessed not so much by taste or appearance, but by the potential additions that may have found their way into it. These included aquatic life forms. In subsequent centuries individuals continued to marvel at the manner by which humans choose to consume a liquid substance that once had living creatures swimming in it. The prominent marine conservation biologist Elliot A. Norse recently reflected that 'in every glass of water we drink, some of the water has already passed through fishes, trees, bacteria, worms in the soil, and many other organisms, including people.' It is for that very reason that some individuals remained sceptical about water drinking. The American comedian and writer W. C. Fields once bluntly announced: 'I never drink water because of the disgusting things that fish do in it.'

It was in the early modern period that certain scientists began striving to resolve concerns such as these by developing technological methods of purification. The period witnessed

Water purification systems, Bret Lake, Switzerland.

significant adjustment in the realm of science, an area of inquiry that was becoming recognizably modern and empirically driven in nature. In the midst of this broader development, scientists undertook important investigations into water conversion. Ingesting many forms of water was clearly unsafe for humans. Yet some scientists were growing weary of this cynical perspective and instead began to exhibit a determination to manipulate and filter waters so that impure varieties of the substance could be transformed into palatable ones.

The renowned English philosopher and scientist Francis Bacon certainly believed in the potential of water filtration. In the early 1600s he set to work developing methods of salt water desalination, a technique designed to remove salt from seawater and render it palatable. Bacon's ideas were enormously ambitious. If they had been achieved, they would have opened up the possibility for human societies to be supplied with a potentially limitless supply of potable water. Although theoretically possible, perfecting the correct procedure was no easy task. Following his experiments, Bacon reported that

> Trials hath been made of salt water passed through earth through ten vessels, one within another, and yet hath not lost its saltiness as to become potable (but when) drayned through twenty vessels hath become fresh.

Despite facing considerable technical difficulties, Bacon's work laid the foundation for further experiments in filtration. The most notable were those pursued by the Italian physician Lucas Antonius Portius, who vastly elaborated upon Bacon's ideas and went on to develop sophisticated techniques employing three pairs of sand filters. In the early 1700s the famed Parisian scientist Philippe de la Hire felt confident enough to present a plan to the French Academy of Sciences in which

u.s. War Department poster, 1944.

he proposed that every Parisian household could have access to a sand filter and a rainwater cistern if they adopted his techniques. Innovative ideas such as these constituted an important shift in mindset. The possibility that all citizens could have access to drinking water was one of potential significance.

Nonetheless the idea that seawater could be transformed into potable water on a large scale remained impractical due to a range of financial and practical limitations. In subsequent centuries scientists continued to be interested in seawater conversion and filtration, yet their efforts were less ambitious. In the late nineteenth century the inventor Alexander Graham Bell decided to establish a way to increase the survival prospects of castaways lost at sea. Working on the precept that moisture could be condensed from the atmosphere, one of Bell's simpler plans required a castaway to float a tumbler on the ocean's surface. If the temperature of the seawater was below the dew point, he believed, moisture would condense in the tumbler. 'Even a dew would be nectar to a thirsty man. But if it is no more, then a dew (adieu) to the man', explained Bell. Interest in seawater conversion also resurfaced during the Second World War when the American government consulted with the Permutit Company, a leading water-conditioning firm, with the intention of inventing a small pack that could be placed in the kits of pilots at risk of being shot down over sea by enemy fire. The proposed kit included a plastic drinking bag that contained a cloth filter at its base and a drinking tube below this. If a briquette were dropped into this bag of seawater, then, after a quick shake, salt could quickly be purged from it.

Salt, however, was not the only undesirable addition that infiltrated water supplies and captured the attention of scientists and inventors. During the Industrial Revolution, enormous factories were built across England and, soon after, other

Western countries. These were primarily designed as sites used for the production of industrial goods, dyes, fuels and cotton. Many of the industrial and technological developments of this period depended upon water power and, later, steam power. Accordingly factories needed to be built near to water supplies. One inevitable drawback was the daily tipping of factory refuse into water sources that had once been relatively clean. In consequence many of the natural waterways surrounding industrial communities became steadily clogged and congested with industrial waste. The cities that expanded around the new industries witnessed rapid population growth and the ejection of high levels of human waste into surrounding rivers and lakes. In addition, as local populations grew dramatically, it was deemed necessary to find some way of supplying them with water.

Few individuals appreciated the magnitude of communally providing safe, distilled drinking water more than the Scottish engineer Robert Thom, who built the first citywide water-filtration plant in Paisley, Scotland. Established in 1804, his innovative plant was able to provide clean filtered water to the entire city, although the relative sophistication of his scheme can perhaps be gauged by the fact that its distribution system consisted of a horse and cart. Subsequently Thom found himself increasingly called upon by town planners and state bodies to offer technical guidance on water provision. His influence was soon felt internationally. In America his sand filtration techniques became popular and were widely heralded for the improvements that they had brought to the smell and taste of drinking water. Clearly social attitudes towards water evolved rapidly in the industrial era, as did those of state bodies, who began to appreciate their important role in protecting, and assuming responsibility for, communal water supplies for the first time.

The water supply of New York, 28 May 1881.

It was in 1828, with the establishment of the Royal Commission on Water Supply in Britain, that the first significant forum came into existence for discussion of the standards of quality that should be expected in public water supplies. At this historically momentous commission, critical questions were posed. Did citizens have a right to good-quality water? Who should regulate supplies? How could safe water be precisely defined and tested? We tend to take these matters for granted today. If unhealthy substances are tipped into a nearby river or stream – let alone our domestic water supply – we presume that a local authority or a water company will eventually dispatch someone to remove it. However, this paternalistic protection is a relatively new phenomenon.

The sophisticated, state-supported water-supply schemes under discussion in the industrial age were intended to replace often haphazard systems of procuring and distributing clean water. Take the example of New York, for instance. In 1667 residents were told that public wells had to be constructed

across the city to resolve water access issues caused by the city's rapid expansion. They were informed that they had to undertake this task themselves and at their own expense. Unsurprisingly this suggestion sparked public antagonism, and the outcome was that only one well was built. Further wells were constructed only when a combination of public funding and state-organized financial subsidies was made available. The threat issued – the enforced sale of the belongings of those who refused to partake in building wells – no doubt provided a further important stimulus. Although still responsible for

Poster on the New York water supply created for the Federal Art Project, 1936.

the upkeep of these wells, those residing in nineteenth-century New York could do little when their water supplies later became polluted from tanneries and slaughterhouses on the banks of the Hudson River as the city's industrial sector grew. As an alternative they came to rely increasingly upon so-called 'tea water men' to procure consumable water. These individuals were barrel owners who vended water in buckets and barrels. In many ways they presaged the popularity of purchasing bottled water, which came into vogue centuries later.

In contrast the members of the Royal Commission on Water Supply in Britain envisaged a more organized and efficiently regulated system of water-supply management. It now seemed apparent that potable water could be, and should be, communally supplied. This was an idea with enormous social, cultural and economic implications. These schemes were to replace makeshift systems of water procurement that too often depended upon walking some distance to the nearest well – often some considerable distance away – or being forced to pay those willing to transport water for the privilege of obtaining a refreshing drink.

It is worth bearing in mind that access to fresh drinking water remained relatively restricted outside urban contexts. In his *Rustic Sounds* Francis Darwin, the son of the famed evolutionist Charles Darwin, recounted of his late nineteenth-century childhood days that

> One sound there was peculiar to Down [House, Darwin's home] – I mean the sounds of drawing water. In that dry chalky country we depended for drinking water on a deep well from which it came up cold and pure in buckets. These were raised by a wire rope wound on a spindle turned by a heavy fly-wheel, and it was the monotonous song of the turning wheel that became so familiar to us.

CITY OF BOSTON.

BOSTON WATER WORKS.

Office of the Superintendent of the E. Division,
No. 221 Federal Street, formerly 21 Sea St.

You are hereby notified that the water will be stopped upon your premises for about five hours, for the purpose of making repairs on the works.

A. STANWOOD,

Supt. of Eastern Division.

N. B. As notices like this must be prepared before they are required for use, it is impossible to state in them the precise time when the water will be stopped, but you are liable to be deprived of it in a few minutes after this has been left at your door.

The importance of preventing the great damage that would sometimes result from a leak, if not properly checked, will explain the cause of an occasional notice with no stoppage of water, and of an occasional stoppage of water without any notice.

If you have a *close* boiler of any kind on your premises, and have *no cistern* or properly constructed *safety valve*, and depend entirely upon the pressure of the Cochituate water to keep your boiler filled, you are hereby warned that you may be in great danger unless the fire under your boiler is extinguished.

Suspension notice, Boston Water Works, 1863.

The well-house, gloomily placed among laurel bushes, had a sort of terrifying attraction for us, and I remember dropping pebbles and waiting – it seemed ages – for them to fall into the water below. We believed the well to be 365 feet deep, also that this was the height of the dome of St Paul's – I have never tested the truth of either statement. The opening was roofed in by a pair of hinged flaps, or doors, and I especially liked the moment when the

rising bucket crashed into the doors from below, throwing them open with a brutal and roistering air which one forgave it as having made a long and dangerous journey up from the distant water.

Water can contain many things. This was well understood in the past. It is little wonder, then, that the quest to convert certain waters into consumable substances, to rid them of their salt content or pollution and to establish ways of supplying safe water to citizens captivated many early modern and nineteenth-century experts. They envisaged a utopian future world in which potable water would be freely available to all with the ameliorative assistance of science and technology. The development of new technologies of water provision also encouraged the formation of new ideas on the rights of citizens to have access to appropriately managed supplies.

The notion that citizens have a right to consumable water remains with us today; in fact, it is often taken for granted in Western societies. However, discussion of the techniques used to purify water has remained clouded by concern over their potential for harm. For instance, the issue of water fluoridization began to generate considerable controversy in the 1950s when opponents and health reformers challenged the ethics, safety and efficacy of the practice. Although many continental European countries decided to cease fluoridizing water, condemning the practice as unsafe, many English-speaking countries still persist in the practice. Those alarmed by the alleged health implications of fluoridization publicly insist that it is directly responsible for serious health problems, including dental fluorosis, a complaint that afflicts the development of teeth during childhood. During the 1950s and '60s, the most vociferous opponents of water fluoridation even went so far as to claim that the practice was a communist plot designed to

undermine public health. The prominent American biochemist Dean Burk added that 'fluoridation is a form of public mass murder.' Evidently the question of how to render water suitable for human consumption is historically rooted, but also continues to spark heated debate.

"At a rough guess I'd say it was water on the knee!"

American comic postcard, mid-20th century.

4
Water and Healthy Bodies

As a food, it appears, from many instances, that water
alone is capable of sustaining human life a much longer
time than could be well imagined. Water, as a common
beverage with all kinds of aliment, affords the best
and most universal diluents in the world. For this
purpose, that which is purest, lightest, softest and
most transparent, is undoubtedly the best.

Anthony Fothergill, 1785

Physicians and scientists have long recognized the health
benefits of drinking water. Even in periods when the safety
of the substance could not always be depended upon and
when it had consequently fallen out of fashion, experts upheld
the physical advantageousness of regular water consumption.
Water assists digestion, cleanses the inner body and promotes
strong blood – so leading physicians and medical authors have
asserted since around the eighteenth century, and have continued to do so ever since.

The Anglo-American poet W. H. Auden once perceptively wrote that 'thousands have lived without love, not one
without water.' Nonetheless, historically many individuals
certainly steered away from consuming water if they could

help it. The English were once internationally renowned for their dislike of the substance, bordering on a phobia. This aversion stemmed in part from a commonly held early modern belief that water consumption predisposed humans to dropsy, an abnormal accumulation of liquid under the skin that causes intense swelling. The first English travellers to reach America reportedly held a deep distaste for drinking water, because they were more accustomed to drinking beer and ale in their native country. Presumably their persistent drunkenness and rowdiness did little to calm the nerves of the native populations upon whose land they were beginning to encroach. Similarly, when the French clergyman and botanist Père Labat was captured by the Spanish army in the 1690s during the Nine Years' War, he reportedly informed the chaplain that 'only invalids and chickens drink water in my country.'

If a European living in an earlier period had been asked what his or her favourite drink was, he or she may well have responded by referring to an alcoholic beverage. Individuals who copiously consumed water found themselves at risk of being socially disregarded. One Frenchwoman, Catherine Beausergaut, warranted a mention in a book entitled *The Cabinet of Curiosities* (1824) because of her seemingly insatiable thirst for water. As an infant, she reportedly drank two pailsful per day and, as an adult, up to 20 pints per day. After her parents banned her from drinking water, Beausergaut clandestinely procured water from rivers, fountains and neighbours. Evidently society had yet to adjust to the idea that water consumption was a normal activity, not a transgressive one. Today Beausergaut's voracious thirst for water would barely pass comment. In earlier centuries those who relished water found themselves at risk of being shunned and avoided, or even dismissed as a social curiosity.

Michele Eugène Cheveul, 'Alcoholism', showing the dangers of alcohol below and the joys of drinking water above, 19th century.

This widespread aversion to drinking water raised an important philosophical conundrum: was man actually designed as a water-drinking organism? After all, physicians could only convince the public to drink water if they were able to convincingly demonstrate its value as an appropriate and natural consumable for humans; a difficult feat in periods hindered by precarious water supplies. The eighteenth-century Austrian diplomat and composer Gottfried van Swieten decided that humans were definitely not natural water drinkers upon observing that 'while girls are daily sipping tepid water liquors, how weak and how flaccid doe they become!' Some decades later, the prominent English physician William Lambe ridiculed those who believed that God had designed man to partake in such an animalistic ritual as drinking water by pointing out that

> Now I see that man has the head elevated above the ground, and to bring the mouth to the earth [to drink water], requires a strained and painful effort. Moreover, the mouth is flat and the nose prominent, circumstances which make the effort still more difficult.

These negative perceptions of water drinking were intellectually underpinned by research pursued centuries earlier by the prominent scientist Isaac Newton, who had hypothesized that, in contrast to all other animals, God had not designed man as a water drinker. Newton connected this intriguing claim to his understanding of man as the only creature that cooked meat prior to consuming it, a process designed to render flesh digestible as well as appetizing. According to Newton, the consumption of hot food generated surplus internal bodily heat which then needed to be quelled with a cold substance if bodily temperature were to be regulated. In Newton's view man

had begun to consume water at some point in his prehistoric past to fulfil this natural necessity for internal cooling.

However, rather than portraying this natural evolution of man into a water-drinking creature in a positive light, Newton insisted that man had in fact significantly deviated from his natural, original state by partaking in a potentially dangerous culinary habit that left him exposed to the potential health problems associated with drinking water. This scepticism persisted for some centuries. Even as late as the mid-nineteenth century, the English Royal Navy officer Frederick Marryat, an acquaintance of Charles Dickens, could be found declaring: 'water, indeed, the only use of water I know is to mix your grog with, and float vessels up and down the world. Why was the sea made salt, but to prevent our drinking too much water?'

Nonetheless, despite these occasional outbursts, the tide of public opinion had started slowly to turn in favour of the practice of drinking water by the Victorian period. This was a result of the increasing recognition and appreciation by communities of scientists and physicians of the benefits to health that we attribute to it today. The idea that drinking water might be a disadvantageous bodily practice seems alien to modern people. Yet just two or three hundred years ago, experts found that they had to work remarkably hard to convince the public of the value of their suggestions.

In the face of widespread public aversion, many medical authorities began to publicly glorify water as a necessary and desirable substance for human consumption. Some went too far. Instead of depicting water as a potentially harmful substance, in his publication *The Curiosities of Common Water* of 1723, British medical author John Smith declared not only that water was healthy but that drinking it could cure all manner of ailments, ranging from consumption and gout to smallpox and even the plague. The psychological benefits of

drinking water also captured Smith's attention. Chronic grief and suicidal desires, it seemed, could be cured by water. As Smith explained:

> Being very hypochondrial, and of a melancholy temper, I have often been strangely dejected in mind when under grief for some misfortunes, which sometimes have been so great, as to threaten danger to life; in which fits of grief I always found the parts within my breast very uneasy, and sometimes continued long; but now I have found a good remedy, for upon drinking a pint or more of cold water, I find ease in two or three minutes, so that no grief seems to afflict.

Most eighteenth-century physicians were not quite as obsessed with water as was Smith, or confident enough in its curative potential to endorse its consumption as a therapeutic magic bullet that could cure all bodily and psychological disorders. Presumably common sense dictated to the majority of physicians that water alleviated, but failed to resolve, the symptoms of serious illnesses. Nonetheless many popular health writers did gradually choose to venerate water not only as a natural drink for man but as one that was *essential* for the healthy maintenance of bodily systems.

Take the views of Dr Daniel Oliver, for instance, a physician who claimed that 'the moment we depart from water, we are left, not to the instinct of nature, but to an artificial taste. Under the guidance of the instinct, God has implanted within us, we are safe, but as soon as we leave it we are in danger.' Or Dr Christopher William Hufeland's pronouncement that

> The best drink is water, a liquid commonly despised, and even by some people considered prejudicial; I will not

Drinking fountain in Prospect Park, Brooklyn, New York, 1870–90.

hesitate, however, to declare it to be one of the greatest means of prolonging life: it is the greatest promoter of digestion, and by its coolness and fixed air, it is an excellent strengthener of the stomach and nerves.

No-one had more faith in the physical and psychological virtues of water than the Austrian peasant farmer Vincenz Priessnitz, who acquired international fame because of his apparent ability to miraculously cure human and animal conditions by wrapping afflicted body parts in parcels containing water. Word of his apparent wonder cure rapidly spread. Before long, Priessnitz found himself compelled to transform his small village cottage into a sanatorium as his popularity grew. His reputation further blossomed in 1826 following a personal invitation to heal Anton Victor, brother of the emperor of Austria. Mass daily pilgrimages to the small village of Gräfenberg soon followed. In 1827 alone Priessnitz received visits from 1,500 patients, including a monarch, a duke, a duchess, 22 princes and 149 counts and countesses.

Although Priessnitz's therapeutic techniques were primarily concerned with applying water to the injured or diseased

Charles Émile Jacque, 'The Hydropath's Second Treatment: Immersion, Submersion and Contortion!', 1880s.

parts of the bodies of those who visited him, he also promoted the regular consumption of plentiful amounts of water and recommended to all of his patients that they consume no fewer than twelve glasses daily. It was common, Priessnitz explained, for patients to feel nauseous upon commencing this course of treatment, but this was owing to the stomach – believed to contain the remains of diseases – having been disturbed by the liquid. Priessnitz also administered injections of water into the eyes, ears and nostrils of some of his less

fortunate patients. Others were advised to sit for hours in cold baths of water, and were even occasionally plunged into these baths should all other treatments be failing. Many of Priessnitz's disciples were renowned, and regularly ridiculed, for consuming up to 40 goblets of water per day and immersing themselves in snow should the opportunity arise.

Although practitioners of orthodox medicine invariably dismissed activities like these as quackery, Priessnitz

'The Water Cure', German lithograph, 19th century.

appreciated, as physicians do today, that water cleanses the stomach and kidneys when ingested. His ideas also tapped into traditional rituals of inner purification still common within traditional medicine. The benefits of water cures are still occasionally upheld by physicians for their bodily value within certain strands of Eastern alternative medicine that reject what they see as a Western over-reliance upon orthodox medication and instead promote water as a natural bodily cleanser.

Priessnitz's ideas had an immediate international influence and encouraged many others to become just as obsessed with water. When Dr Charles Shieferdecker inaugurated his Hydriatic Institution in Philadelphia in the mid-nineteenth century, he heralded it as a therapeutic site where diseases ranging from pox to inflammation of the brain could be cured. 'I am convinced that cold water, exercise, a proper diet and pure air will give men the age of 150 to 200 years', asserted Shieferdecker, adding that 'accident ought to end the life of

The Round Hill Water Cure Institution, Northampton, Massachusetts.

a human being, not disease.' Contemporaries regularly cited examples of humans living for up to 30 days on nothing but cold water and depicted the human system as a hydraulic machine composed of hundreds of thousands of tubes designed to convey liquid around the body.

Spa treatment was a further medical craze that created divided opinion among orthodox and non-orthodox medical practitioners. The nineteenth century witnessed a fashion for spas and mineral water across many Western countries. The emergence of technological innovations, including the railway, opened up new possibilities for individuals to travel to previously remote spa resorts. Contemporaneously scientists developed a strong interest in analysing and testing waters found at spas. Dr Augustus Granville, in his popular *Spas of England* of 1841, meticulously analysed the mineral properties of English wells and helped to popularize the idea that mineral waters served medicinal purposes. Unlike earlier accounts of spas, Granville's was free of superstition and did not falsely promote the use of any particular spa to attract custom. Across Europe individuals – mostly drawn from the middle and upper classes – flocked to the now famous spas of European cities, including Bath and Wiesbaden, in the hope of being cured of their ailments by bathing in and consuming mineral-rich waters. Spas were also popular throughout many late nineteenth-century British colonies and served as potent reminders of home for colonizers. Imperial expansion also offered opportunities for scientists to compare the chemical composition of waters found internationally with those at home and to widen their understanding of the substance.

Popular water cures were also developed in relation to key spa sites. The Glen Haven water cure, widely circulated in the 1840s, made the following recommendations:

Water should be taken as fresh as possible, so as to secure the benefits of the carbonic acid gas that it contains. Water should be drunk in the morning before breakfast, according to the capability of the patient. As a general rule for beginners in the cure, from one half a glass to three glasses before breakfast. About midway between meals, both in the fore and afternoons, the same quantity as before mentioned should be drank. No water should be drank except to quench thirst, under two hours after eating; and but moderate quantities during meals.

Some contemporaries remained sceptical about the rampant enthusiasm for water spearheaded by Priessnitz, Granville and others. At worst water drinkers found themselves exposed to not entirely unpersuasive accusations of obsessiveness and health radicalism. This was evident in the French poet Charles Baudelaire's statement that 'a man who drinks only water has

Postcard of the German spa town of Aachen, 1910.

Postcard for the Kochbrunnen hot spring spa resort, Wiesbaden, Germany.

a secret to hide from his fellow men', an explicit reference to an alleged fear of losing oneself or admitting one's secrets while consuming alcohol. Despite the existence of cynicism, the ideas forged in the eighteenth and nineteenth centuries about the healthiness of drinking water have endured to this day.

An important shift in attitudes had clearly occurred since the Middle Ages. For nineteenth-century enthusiasts, water sustained health; it was essential and could be imbued with miraculous healing and curative properties. The foundations of the modern understanding of water as an essential component of bodily health can therefore be considered to have been established in this period, albeit in a sometimes over-enthusiastic manner. However, medical and scientific communities still had considerable work ahead of them in convincing the consuming public that drinking water was an activity that did not compromise health or, at worst, result in death.

5
Making Water Safe

> Distill, or filter water as often as you please, and it will
> nevertheless in time turn putrid in the sun, and by its bubbles,
> scum, sediment and taste, afford evidence of its impurity.
>
> *New Monthly Magazine* (1823)

When the late eighteenth-century Dutch tradesman and scientist Antoine van Leeuwenhoek peered into his microscope to determine what was inside a sample of water, he was surprised to find a disturbing microcosmic realm of strange-looking, minute organisms peering back at him, which had previously been hidden to the human eye. In his lifetime Leeuwenhoek enthusiastically invented over 500 simple microscopes and conducted an array of important experiments with them. He was therefore well positioned to be one of the first individuals able to expose water as a liquid sometimes, if not typically, teeming with microscopial life. Whether the tiny, mysterious organisms discovered in his water samples were potentially harmful to the human body seemed somewhat unclear to Leeuwhenhoek. Yet the presence of these ugly, alien-looking creatures hardly seemed desirable or pleasant.

Prior to Leeuwenhoek's discoveries, water quality tended to be appreciated in terms of where a water sample had been

obtained from. Various 'types' of water were recognized. The eighteenth-century physician Thomas Percival wrote that

> In Minorca, where the water which the springs and rivulets afford, is often brackish, and always hard, obstructions, indurations, and swellings of the abdominal viscera, together with flatulency and indigestion, are the most common diseases to which the inhabitants are subject. And it is remarkable that large spleens and tumefied livers are not peculiar there to the human species, but are incident also to brutes; especially to the sheep.

For Percival, it was the type of water and its initial geographical location that determined its consistency and bodily impact. The significance of Leeuwhenhoek's findings was that they added important new dimensions to the myriad ways in which water could be understood. His ideas implied that the quality of all water, regardless of its origin, could be measured against a gradated system of purity concerned with the number of microbacterial organisms contained within it and how deadly or dangerous these would be if ingested.

As Western countries industrialized, fears about the potentially unsavoury bacterial content of water intensified. In the 1830s the English writer and cleric Sydney Smith wrote to Countess Grey warning her that 'he who drinks a tumbler of London water has literally in his stomach more animated beings than there are men, women, and children on the face of the globe.' Similarly in the 1850s the English physician, chemist, microscopist and leading food reformer Arthur Hill Hassall sparked a mid-Victorian public panic over the content of communal drinking water supplies. In his widely read *Microscopical Examination of the Water Supplied to the Inhabitants of London and Suburban Districts* of 1850, as well as in the pages

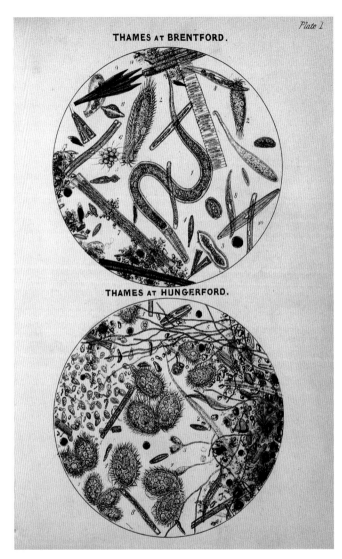

Thames at Brentford and Thames at Hungerford', from Arthur Hill Hassall, *A Microscopic Examination of Water Supplied to the Inhabitants of London and the Surburban Districts* (1850).

of the infamously outspoken radical medical journal *The Lancet*, Hassall printed sketches of the disgusting organisms that, using his microscope, he had found residing in London's drinking water supplies.

Hassall's sketches were somewhat misleading, as they created the false impression that all of the organisms ominously staring out of the page at his readers could be observed in one single glance into a microscope. In fact Hassall was later forced to admit publicly that his water samples had not been quite as crowded as his illustrations suggested. Despite this Hassall's key achievement was to popularize – indeed cement – the public understanding of water as a substance potentially swarming with microscopic life forms, some of which would be harmful if consumed. His emotive images were at the vanguard of campaigns for water reform. English politicians routinely referred to them in Parliament, while contributors to medical and literary periodicals debated them internationally. Sceptics, normally from water companies, insisted that no direct evidence existed to confirm the harmfulness of swallowing these minute organisms. Yet both common sense and scientific reasoning dictated that these mysterious organisms were hardly appetizing and that at least some of them were probably harmful.

The understanding of disease transmission evolved significantly. It had always been understood that boiling water somehow helped to make water safer for purposes such as cooking. Yet it was only in the nineteenth century that scientists could for the first time accurately pinpoint ingested bacteria and germs as directly responsible for the onset of many waterborne human diseases. This conceptual development revolutionized understandings and definitions of water purity. Large clusters of citizens residing close to one another in the ever-expanding industrial towns and cities meant that waterborne

epidemic diseases, including cholera and typhoid fever, could spread more quickly than ever before. Alarmingly these diseases had a high potential to kill or permanently disfigure large, concentrated masses of people at any given time. In the absence of a bacteriological framework for understanding disease, in 1831 alone 6,536 people had perished from cholera in London (and 55,000 nationwide), while 20,000 had succumbed in Paris (with an additional 80,000 deaths recorded nationally). In the same year hundreds of thousands of individuals had also died from cholera in Russia, Hungary, America, Canada and Egypt. Typhoid was a further waterborne killer that claimed the life of Queen Victoria's beloved husband, Albert, in 1861.

Initially it proved difficult for some to contemplate how water, even when it appeared to be pure, transparent and clean, might contain deadly microbacterial content. When the prominent English physician and anaesthetist John Snow suggested that polluted water transmitted cholera, few believed him. Diseases were transmitted by bad smells and unsavoury air, his critics insisted. However, following an epidemic that killed almost 11,000 Londoners in 1854, Snow noticed that higher incidence levels had occurred among those who drew drinking water from a public well located near his Soho surgery. Meanwhile workers from a nearby brewery, who were fortunate enough to have their own water supply, remained unaffected. Further investigation demonstrated that a nearby leaking sewer had contaminated the communal drinking water supply and was directly responsible for the spread of cholera. Snow's centrality to the history of drinking water is beyond doubt. After connecting cholera to contaminated water, he went on to develop novel ways of using chlorine to kill cholera bacteria residing in water and took steps to discount the idea that if water was good-tasting and odourless, it was safe. Ultimately

Outside water supply, Washington, DC, July 1934.

Drinking fountain in Charleston, South Carolina.

Snow's ideas influenced a generation of reformers who sought to cleanse public water supplies.

As Snow's discoveries came to be gradually accepted, local authorities from several British cities took steps to treat all communal drinking water supplies with sand filters and chlorine. Some claimed to have done so, but evidently had not. Investigations into a cholera epidemic that struck Whitechapel, London, in 1866 produced some unexpected findings. One local resident recollected that

Our water pipe was stopped up, and as I am a plumber I cut the pipe, and to my astonishment found a dead eel, nine inches in length, which must have come through the above works and which is quite demonstrative that the water is not filtered.

The company under fire was the East London Water Company, whose water supplies had become heavily polluted. In an era when citizens were beginning to demand and expect clean, trustworthy water supplies, the discovery of untreated water contaminated with sewage and even large living animals now seemed scandalous and irresponsible.

Further technological developments emerged as medical scientists and physicians connected insanitary water to the onset of a spectrum of other potentially deadly diseases. Two American men, Halsey Willard Taylor and Luther Haws, invented the modern drinking fountain in the early 1900s. Taylor's father had died of typhoid fever after consuming water from a contaminated public supply. This unfortunate incident motivated Taylor to invent a new type of water fountain designed to ensure safe drinking water. Haws, meanwhile, was a part-time plumber, sheet metal contractor and sanitary inspector for the city of Berkeley, California. After observing children drinking out of a communal cup, he feared that the manner by which the public shared their water supplies posed a distinct health hazard. Haws went on to invent the first faucet designed for drinking.

It is important to note that the management of water-borne diseases occasionally prompted controversy. This was particularly the case in scenarios in which Western medical scientists and state bodies intervened in non-Western regions. Cholera continued to afflict less affluent communities, in countries including colonial India, during the late nineteenth century.

One way in which the British colonial state attempted to stem the spread of cholera was to draw links between traditional Hindu pilgrimages and cholera. Much to the chagrin of Hindu communities, British state officials took steps to sanitize traditional pilgrimage sites and to effect greater control over the movement of pilgrims to the River Ganges. India was derogatively considered in the West as the home of cholera. Accordingly quarantines were set in place against Indian vessels carrying pilgrims. Steps such as these also controversially disrupted the traditional pilgrimage of Indian Muslims to Mecca.

Despite difficulties such as these, the rapid improvements witnessed in standards of drinking water stand as a testimony to the achievements of Victorian medical science, at least in

Drinking fountain in Saint-Paul de Vence, France.

Western society. By 1896 cholera had become so rare in many Western countries that physicians had begun to classify it as an exotic disease. However, to this day it has continued to plague cities with inadequate sewer systems across South America, India and Bangladesh.

Evidently shifting understandings of disease aetiology, prevention and management held important implications for the ways in which water was historically consumed. The ideas and investigative strategies forged in the late nineteenth century significantly developed understandings of the potential content of drinking water. In earlier periods contemporaries pondered over matters including how to remove salt from water. Although the presence of the remains of fish and other sea creatures occasionally perturbed them, few stopped to consider the possibility that water teemed with life invisible to the human eye that was responsible for sporadic outbreaks of epidemic diseases and precarious communal health. It was only when water's dangerous potential was elucidated that serious efforts could be made to transform water into something safe for all to consume, and that, simultaneously, new demands were placed on those supplying water to regulate their supplies.

6
Alcohol or Water?

During the eighteenth century, Blash de Manfre – or 'the water spouter', as he was more commonly known – acquired international fame and notoriety by travelling around Europe conducting an impressive magic trick. Upon swallowing a large quantity of drinking water, De Manfre would then discharge the liquid from his stomach in a new form: as fine wine, beer, oil or milk. The authenticity of his ability to transform water in this memorable way seems somewhat dubious, but his reputation blossomed to such an extent that he regularly found himself called upon to perform before captivated audiences that sometimes included European emperors and kings.

De Manfre's antics would have horrified nineteenth-century temperance campaigners. In the industrial age alcohol was increasingly demonized by the remarkably proactive temperance movement, whose members passionately preached – mostly to the industrial working classes – about the health virtues of consuming non-alcoholic substances and the comparative evils of alcoholic intoxication. Unsurprisingly given the comparative reliability of distilled, processed alcoholic beverages, considerable effort was needed to convince the public that water consumption was natural and healthy and that reliance upon other liquid substances, especially alcoholic ones,

Royal Crescent, Bath.

could be physically harmful. Temperance advocates and water enthusiasts strove firmly to establish in the public mind a sense that water was the only natural drink that man had been designed to consume. The effectiveness of this strategy depended upon condemning all others potable substances as man-made and 'artificial'.

Contemporary perspectives on the natural bodily benefits of water consumption were summed up in 1823 by an anonymous contributor to the *New Monthly Magazine* who suggested that

> As water can perform such great things, and at the same time, because it has no taste, it neither stimulates the appetite to excess, nor can produce any perceptible effect on the nerves, it is admirably adapted for diet, and we ought, perhaps, by right, to make it our sole beverage, as it was with the first of mankind, and still is with all the animals.

However, advocates of perspectives such as this faced a scenario in which high alcohol consumption was a well-established social practice in many Western countries. Until the First World War public houses in London opened as early as five o'clock in the morning, allowing workers to enjoy a pint or two of strong British ale before commencing work.

Temperance advocates also had to grapple with the thorny philosophical question of whether God had actually created man as a water-drinking creature. Indeed, prior to the industrial period, various prominent individuals had agreed that alcoholic beverages, including wine, were in fact a blessing that had rescued mankind from the precarious practice of consuming water. Benjamin Franklin, scientist and founding father of the United States, had argued in the eighteenth century that

> Before Noah, man having only water to drink, could not find the truth. Accordingly . . . they became abominably wicked, and they were justly exterminated by the water they loved to drink. This good man, Noah, having seen that all his contemporaries had perished by this unpleasant drink, took a dislike to it; and God, to relieve his dryness, created the vine and revealed to him the art of making *le vin*. By the aid of this liquid he unveiled more and more truth.

The influential Enlightenment philosopher and physician John Locke had even recommended that young children should be fed beer instead of water to promote normal childhood growth and health. Franklin and Locke's views were expressions of a pervasive cultural ethos that privileged light alcoholic consumption over water drinking.

However, water was undeniably cheaper, less addictive and healthier than alcohol despite its potential to be impure

SPRING

John C. Sinclair, print of a drinking fountain in Fairmount Park,
Philadelphia, *c.* 1870.

and contaminated. Moreover water could more easily be construed as a natural beverage in comparison to alcoholic drinks that required processing. Alcohol may instinctively have seemed safer than water but its widespread consumption carried its own set of problems. In addition to inducing a lack of coordination and merry behaviour, alcohol also prompted violent tendencies and chronic poverty should family budgets be squandered in the public house.

For reasons such as these, water drinking enthusiasts decided to challenge long-established views on the primacy of alcohol by insisting that water, not wine, was in fact God's gift to mankind. The idea that water was a more appropriate and natural item of human consumption than alcohol prompted the nineteenth-century French novelist Victor Hugo to proclaim that 'God made only water, but man made wine.' Those who abstained from alcohol, and who were often horrified by the social fallout of cultures of alcohol consumption, chose instead to swear by the health benefits of water. The nineteenth-century radical journalist William Cobbett, for instance, once proclaimed that

> In the midst of a society, where wine and spirit are considered as of little more value than water, I have lived two years without either; and with no drink but water, except when I have found it convenient to obtain milk; not an hour's illness; not a headache for an hour; nor the smallest ailment; not a restless night; not a drowsy morning have I known during these two famous years of my life.

Similarly the prominent nineteenth-century Irish-born physician and medical author James Johnson felt so confident about water's physical benefits that he declared:

The water drinker glides tranquilly through life without much exhilaration, or depression, and escapes many diseases to which he would otherwise be subject. The wine drinker experiences short but vivid periods of rapture and long intervals of gloom; he is also more subject to disease. The balance of enjoyment then, turns decidedly in favour of the water drinker.

The rising repute of water as a potable substance in many nineteenth-century Western societies can be partly ascribed to the rigorous efforts made by temperance campaigners to wean citizens away from their beloved alcohol, and, in turn, stem the social disruption that they linked to working-class alcohol consumption. Tellingly when the British MP and philanthropist Edward Thomas Wakefield penned his *Plea for Free Drinking Fountains in the Metropolis* in 1859, he hoped that the provision of free drinking water supplies in major towns and cities would promote not only health but temperance and morality. Wakefield's plans were effective. By the time his Metropolitan Drinking Fountain Association movement was established in Liverpool during the 1850s, a burgeoning understanding of water's bacterial content was tarnishing the reputation of many British water companies. Some of these had fallen so far out of public favour that local government bodies had felt compelled to gradually buy them out. Upon securing greater control over public water supplies, many local authorities set to work on an impressive construction programme designed to widen public access to bathing facilities and communal drinking water fountains. Many water fountains were positioned close to public houses, offering a potent visual reminder to pub-goers of the healthier and more virtuous path that could be followed. Although now less concerned with banishing alcohol, the Drinking Fountain Association still thrives in Britain, where

it relies upon grants from organizations such as the National Lottery to restore existing fountains and build new ones.

The ever-increasing safety of water drinking strengthened the case for temperance. As access to reliable supplies expanded, the rationale behind the argument that alcohol needed to be consumed because of water's potential danger seemed increasingly obsolete. This perspective appeared increasingly outdated in an era marked by effective efforts to publicly provide potable water. Water was also venerated as a healthy alternative to alcohol across the Atlantic Ocean, although American temperance activity suffered from some initial teething problems. In 1835 the author Henry Cook Todd recalled in his *Notes upon Canada and the United States* that sixteen water drinkers had dropped down dead on one particularly hot summer day alone, soon after teetotalism had arrived in New York. It later transpired that these unfortunate individuals had been actively encouraged to stop their usual practice of diluting water with alcohol. Their deaths were subsequently blamed upon the poor quality of New York's water supply, an outcome that did much to discredit the city's active water-drinking campaigns.

Temperance did, however, gradually take hold in America. The commonly used term 'off the wagon' originated there: it is a direct reference to someone who has fallen 'off the water wagon' – or the water carts then used to hose down roads and keep off dust. If someone was 'on the wagon', they would climb on a water cart to quench their thirst rather than retiring to a public house. American anti-alcohol campaigners revered water to such an extent that temperance communities there became colloquially known as the Cold Water Army. Striving to transform communal drinking habits, the 'army', emulating their European equivalents, upheld water as a natural, healthy alternative to ale and beer and famously marched across major

towns and cities singing songs on the subject to exemplify their point. Take their 'Song for Independence Day' (to be sung to the tune of 'Yankee Doodle'), which ran as follows:

Cold water is the drink for me,
Of all the drinks, the best sir;
Your grog, of whate'er name it be
I dare not for to taste, sir.

Give me dame nature's only drink,
And I can make it do, sir;
Then what care I what other think, –
The best that ever grew, sir.

Your artificial drinks are made,
The appetite to please, sir,
And help along the honest trade,
Of those who live at ease, sir.

Your logwood wine is very fine,
I think they call it 'Port,' sir;
You'll know it by this certain sign,
Its roughness in the throat, sir.

'Tis true that yankees are most shrewd,
And wooden nutmegs make, sir;
But who'd have thought Port wine was brew'd
This side the big salt lake, sir.

We need not send to Portugal,
Nor go to good old Spain, sir;
The best of wine is at our call,
Port, Lisbon, or Champaigne, sir.

They'll make us any kind we choose,
Without the aid of grape, sir;
And when 'tis done, will not refuse
A price to make it take, sir.

Some love to swig New England rum,
And some do Cider choose, sir;
But, so they only make 'drunk come,'
No matter what they use, sir.

But I'll not touch the poisonous stuff,
Since all the brooks are free, sir;
Give me cold water, 'tis enough,
That cannot injure me, sir.

Leading American water drinking advocates included Sylvester Graham and John Harvey Kellogg, health reformers whose names remain familiar today because of the popularity of the healthy cereals that they invented and promoted. In America anti-alcohol sentiment ultimately proved so influential that it was partially passed into legislation during what is now referred to as the period of Prohibition between 1919 and 1933, an era when strong alcoholic beverages were banned across the nation. W. C. Fields was notably unimpressed. On one occasion, he sardonically declared that 'Once, during Prohibition, I was forced to live for days on nothing but food and water.'

It might be expected that the problematic issue of the Bible being replete with stories involving the consumption of wine and alcohol would have presented a considerable obstacle to the temperance movement, especially given that many temperance societies self-consciously associated themselves with Christianity and fashioned their arguments with

direct reference to scripture. Yet the prevalence of references to alcohol scattered throughout the Bible failed to deter them. Temperance authors such as Benjamin Parsons effectively rewrote history by asserting that ancient civilizations mostly consumed water. Glossing over multiples accounts of drunken Roman orgies and biblical stories of Jesus turning water into wine, Parsons insisted that when alcohol was consumed, it was done so only occasionally and even then only at religious festivals. Evidently Parsons consciously adopted an agenda of distinguishing between alcohol consumption as a way of life (as, in his view, was too often the case in the nineteenth century), and as an occasional pleasure, by revealing how the healthiest empires of the past had subsisted upon water alone; a fact that accounted for their strength and vitality.

Referring to ancient empires in a period of expanding imperialism and empire-building was an effective and emotive rhetorical strategy. In addition anti-alcohol sentiment exerted a notable influence in colonial settings. When Richard F. Logan visited the region now known as South Africa in 1874, he critically noted that 'the only disease known at the Bay is

Plastic bottles on conveyor and water bottling machine.

intoxication. Drinking water is brought from Cape Town, and they will give you a bottle of English ale worth twenty-five cents sooner than a drink of water.'

It is difficult, if not impossible, to determine the extent to which the rising popularity of water drinking during the nineteenth century can be accredited to the social pressure directed at citizens to sacrifice their alcohol. However, the rapid shift in social attitudes towards water that occurred was undeniably remarkable. The temperance movement was in many ways fortunate in that its campaign reached a peak at the very same time that state bodies were beginning to actively involve themselves in programmes of improving communal water supplies, and when many influential medical scientists were preoccupied with forging new understandings of the bacterial content of water and elucidating new means of avoiding deaths linked to the consumption of unsafe waters. Without these developments it seems less likely that the public would have increasingly chosen to drink water, even if they had become convinced by claims of its healthy properties. Nonetheless, even today some remain sceptical. As David Auerbach announced in 2002, 'in wine there is wisdom, in beer there is strength, in water there is bacteria.'

7
Marketing Drinking Water

It is curious that so many of us willingly purchase a product that is not only natural but flows freely from our taps. As the popularity of bottled water surged in the late twentieth century, critics insisted that drink-manufacturing companies were attempting to con the consuming public. The American satirist George Carlin sardonically asked: 'Ever wonder about those people who spend $2 apiece on those little bottles of Evian water? Try spelling Evian backward.' Nonetheless the popularity of bottled mineral water, as well as sparkling water, was to become phenomenal. But how did the idea of bottling and vending water evolve?

For centuries the question of how to transport fresh water remained unanswered. Sea journeys made thirst particularly frustrating, as thirsty sailors were taunted by being surrounded by seawater that would only intensify thirst if consumed. Nowhere was the temptation to convert water into something palatable more pronounced than on a lengthy sea journey. Nowhere, it seemed, was water more omnipresent. The English Romantic poet Samuel Taylor Coleridge memorialized this predicament in his poem 'The Rime of the Ancient Mariner':

Water, water, everywhere
And all the boards did shrink
Water, water, everywhere
Nor any drop to drink

At the very same time that European scientists were per-
forming experiments with sand filtration, ships were being
built of such a magnitude that explorers could now safely, but
painstakingly slowly, travel to distant parts of the globe. These
long, tedious journeys raised pertinent questions about how
to keep food fresh on voyages that lasted for months with
few stop-off points. And, perhaps more importantly, how
could drinking water be procured and stored for these lengthy
travels? When Captain Cook undertook his famous worldwide voyages in the eighteenth century, he procured fresh
water by melting the ice that he found floating by him in the
sea. He described it as 'soft and wholesome'. It was well known
that frozen water was divested of its saltiness and was therefore palatable when melted.

However, melting seawater was unfeasible for travellers
traversing warm, non-Arctic regions. In the eighteenth century
the issue of rendering seawater fit for mariners on long voyages
attracted the attention of the British Parliament, who offered
liberal premiums to experimental chemists for the development
of seawater conversion techniques. This financial incentive
encouraged numerous scientists to conduct experiments, the
most successful being those of Dr Charles Irving who, in 1770,
developed a practical technique of distilling water in sophisticated kettles that enabled salt to be filtered from seawater.

In the same century producers began to bottle and sell
mineral water. In doing so they helped to transform water
into a consumable commodity; something that could be
bottled, transported and sold. Bottles have since continued

Budiš, the best-selling mineral water in Slovakia.

to offer the most convenient way of storing and selling water. Efforts to can water were developed during the twentieth century, with the expectation that metal would fare better than bottles in the event of the outbreak of nuclear war. Yet the popularity of bottled water products proved persistent. Mineral water is known to have been collected and vended in bottles

Bottled Evian natural spring water on display at the u.s. Open, New York, 2012.

in Boston from at least the late eighteenth century. When new glass technologies cheapened the cost of producing bottles in the early nineteenth century, the mass sale of bottled water products became both feasible and profitable. Numerous successful entrepreneurial ventures emerged internationally. For instance, the famous Saratoga mineral waters of New York began to be bottled and sold as early as 1809. Impressively by 1856 over 7 million bottles were being produced annually there, and being sold for up to $1.75 per pint.

Most of the recognizable names now attached to the mineral-water industry began to operate during the nineteenth century as a result of new techniques of bottle-manufacturing having been developed and new commercial possibilities for water vendors having subsequently opened up. The Evian bottled water company began life when the Marquis of Lessert first tasted water from a spring in the French town of Evian. After some weeks of drinking regularly from that spring, he noticed that the pain caused by his chronic kidney and liver problems had been significantly ameliorated. Soon afterwards local doctors began prescribing Evian water as a health

remedy. Sensing a unique and potentially profitable marketing opportunity, the owner of the spring closed off public access and began to sell its waters commercially. In 1829 the Société des Eaux Minérales was founded, and Evian water became sold as a drinkable commodity rather than a medicinal substance.

Bottled Perrier water also originated in nineteenth-century France. The spring from which Perrier water is collected is naturally carbonated – or infused with carbon dioxide gas – although entrepreneurs found ways to inject gas back into the product during the bottling process to sustain the original, natural taste and texture of the Perrier spring. The spring, originally called Les Bouillens, had been used as a spa

Limited-edition Evian bottle by Diane von Furstenberg, 2012.

Perrier is a French brand of bottled mineral water drawn from a spring in Vergèze in the Gard département.

since Roman times and water had been bottled and sold in Les Bouillens since 1863. In 1898 the entrepreneur and physician Louis-Eugène Perrier bought the spa and eventually sold it to John Harmsworth, the younger brother of the founder of the *London Daily Mail*, who renamed the water after

Dr Perrier. Perrier was initially advertised as a mixer for whisky, although it later became known as the champagne of mineral waters. The product continues to be marketed as a fashionable high-end water today.

Similarly Badoit water is drawn from the Saint-Galmier spring in the south of France. Initially prescribed by local doctors, it was transformed into a commercial consumable product by Auguste Badoit in 1841. Known for its exhilarating effects, the product was advertised as encouraging cheerfulness and conviviality. Its popularity persisted into the twentieth century, when it became sold widely in French restaurants. By 1958 the Badoit company was producing 37 million bottles per year.

The commercial marketing of water was not restricted to France. San Pellegrino mineral water, from Italy, has been consumed for health purposes since at least the fourteenth century. Leonardo da Vinci is rumoured to have visited the town in 1509 with the specific intention of sampling its water. He later penned a treatise on the properties of the water, which he had enthusiastically tasted. Towards the end of the nineteenth century commercialists began bottling San Pellegrino water and selling it widely not only in Europe but throughout Asia, Africa and the Americas.

Physicians have also commonly associated the popular German water Staatl. Fachingen with well-being, health and mental performance. The German poet and author Johann Wolfgang von Goethe wrote in 1817: 'I wish to be served with Fachinger water and white wine, one for the liberation of the spirit, the other for the animation of it.' Fachingen was yet another water product that was transformed from a medicinal to commercial product. In 1939 Adolf Hitler dismissed his personal bodyguard of five years, Karl Wilhelm Krause, on the basis that he had forgotten to bring a bottle of the water

to a front-line inspection and then surreptitiously attempted to serve him local Polish water, much to Hitler's distaste.

Borjomi water from Georgia was noted for its healing powers before being commercially exploited. In 1829 Russian soldiers first discovered the mineral springs on the bank of the Borjomi River. Colonel Pavel Popov, commander of the regiment, ordered his men to cleanse the springs, bottle samples of its water and transport them to a nearby military base. Popov quickly realized that the spring water soothed his crippling stomach complaints. He ordered rock walls to be built around the spring in addition to a bathhouse. The site rapidly acquired

Badoit mineral water, obtained from natural sources at Saint-Galmier, France.

international fame for its healing effects. A mineral water park was opened at Borjomi in 1850. In 1854 the Russian state commissioned the construction of the first bottling plant in the region.

The legacy of the very simple but highly effective idea of bottling and selling water remains with us today and has had an immeasurable impact on the ways in which drinking water is procured and consumed. The bottled water industry has mushroomed in recent decades, raising billions of dollars annually. Purchasing bottled drinking water fell out of fashion once access to safe public and domestic water supplies became the norm in Western societies during the early twentieth century. That was until 1977, when Perrier famously launched a rigorous advertising campaign across America featuring the famed actor Orson Welles. Spring water quickly evolved from a niche product to a once-again fashionable accessory with broad public appeal. The company's president, Gustave Leven, also publicly promoted its health virtues, making the case for its positive effect on bodily digestion, just as physicians had done 200 years earlier.

Even as recently as the 1970s few people expected bottled water to become quite as popular as it proves to be today. Global purchasing levels have rocketed, especially since the early 1990s. Mineral water has always differed from tap water in the sense that it is recognizably pure. Mineral water is groundwater that has emerged from the ground and travelled over rocks. The only treatment involved prior to bottling is the removal of iron and sulphur compounds and filtration. Among the best-selling and most renowned mineral waters sold globally today are Evian, Perrier and Volvic (all from France), FIJI Natural Artesian (Fiji), Gerolsteiner (Germany), Ferrarelle and San Pellegrino (both from Italy), Mountain Valley (United States), Tŷ Nant (Wales) and Icelandic Glacial (Iceland). The

most expensive bottle of water ever sold was a special-edition Fernando Altamirano bottle designed by the Italian artist Clemente Modigliani, enveloped in 24-carat gold, which contained water gathered from springs across the globe. The $60,000 spent for the bottle was donated to organizations campaigning against global warming.

Occasional controversies have surfaced over the potential of contamination from the plastics used to bottle water as well as water being collected from sources other than the spring. The public tends to believe that bottled mineral water is healthier than tap water, as well as appreciating the convenience of being able to purchase it while on the go. But which tastes better according to the public? Several recent studies have compared bottled and tap water in a quest to determine which option is superior in terms of quality and taste. A debate has ensued in recent decades over the merits of both types of water. Some have observed that the quality of tap water is more rigorously controlled and frequently analysed; others have insisted that bottled water is submitted to more advanced forms of treatment and is therefore less liable to contamination. To a certain extent purchasing choices have always been determined by local circumstances. Water can have a very different taste depending on the region in which it is found. Parisians, for instance, have often claimed that the city's water tastes differently according to the arrondissement they are in. Dissatisfaction with the taste, odour or sight of tapped water in certain regions has persistently influenced preferences for either tapped or bottled water. In New York, for instance, tapped water has tended to be preferred. Yet bottled water is favoured throughout America as a whole. One place seems unconvinced by the benefits of bottled water: Bundanoon, a small town in New South Wales, Australia, where its sale is outlawed at present. Residents of Bundanoon made this decision

after reaching agreement on the damaging environmental effects of bottled water, including litter, and in opposition to a proposal to open an unsightly bottling plant in the small town.

Despite this exception, the ways in which we purchase and consume water have changed dramatically in recent decades. Water has transformed into a commodity, and one that the public have proven increasingly more eager to purchase even despite the availability of water at home in Western societies. In recent years bottled water companies have started to investigate the potential for marketing their products in Asia and Latin America because of the poor quality of potable water in many non-Western countries. Although critics have argued that commercializing bottled water in these countries might distract from finding a more sustainable solution to a relative absence of publicly managed municipal water services, the availability of bottled water products might undeniably resolve some problems in these regions. Nonetheless, in countries such as South Africa, which has established a bottling industry in recent decades, consumption levels pale in comparison to its equivalent in Western countries.

Mass production of lemonade began in the 19th century, although it had been drunk for some centuries previously.

8
Making Water Interesting

Water's bland, unappealing consistency often presented an obstacle for those seeking to encourage its consumption, despite the high visibility of health information geared towards promoting water drinking. In response, concerted efforts have been made to transform water into a more exciting and interesting product throughout the last two centuries. One of the earliest solutions was to make water fizzy. One explanation for the success of nineteenth-century temperance advocates in popularizing water drinking was the evolution of techniques that manipulated water to alter its taste, texture and appearance.

Temperance advocates appreciated that aerating water transformed the liquid into a more palatable and interesting product. Aerated, or carbonated, water is essentially distilled water to which purified air has been added to adjust its flavour and consistency. In the eighteenth century physicians and scientists mostly agreed upon the healthy properties of naturally carbonated water that could be obtained from volcanic springs. Yet capturing and transporting these waters was then a costly undertaking. Accordingly mineral waters tended to be sold only in pharmacies and even then only on a small scale. However, they proved particularly popular among those who could afford them, as effervescent waters soothed stomachs and relieved

indigestion, a complaint that plagued many industrializing societies such as Britain.

In the meantime scientists had begun to inquire into why gas bubbles dissolved in water and to develop methods to artificially recreate that effect. The prominent English chemist Joseph Priestley essentially invented fizzy water by devising methods of infusing water with carbon dioxide. He published his results in 1772 in a seminal scientific research paper entitled 'Impregnating Water with Fixed Air'. Contemporaneously the Swedish chemistry professor Torbern Bergman invented a similar process. The fruits of these scientific endeavours quickly impacted upon commercial behaviour and significantly impacted on the ways in which Western citizens consumed water. What became known as the 'Geneva apparatus' was developed in 1788, a form of producing artificial mineral water whereby carbon dioxide was produced by stimulating a chemical reaction between sulphur and bicarbonate of soda. The gas was then piped into a gas meter and then onwards into a mixing vessel filled with water. In the 1820s the German doctor Friedrich Adolf August Strove improved and perfected this process, and opened a highly successful commercial factory in Dresden. His artificial mineral waters convincingly imitated the mineral waters passing through the Niederselters springs near Hesse, Germany, known locally as seltzers. Strove also developed effective procedures for producing cheap soda water.

Despite the long-standing interest in commercializing the Niederselters spring, the term 'seltzer water', like 'sparkling water', is surprisingly modern, having only gained favour from the 1950s. Although it tends to be used as a generic trade name, it originates from nineteenth-century efforts to artificially mimic waters found in Niederselters, such as that undertaken by cookery author Mrs Marcet, who described seltzer water in 1813:

By analysing that which is produced by nature, it was found to contain scarcely anything more than common water impregnated with a certain proportion of carbonic acid gas. We are, therefore, able to imitate it, by mixing those proportions of water and carbonic acid. Here, my dear, is an instance, in which, by a chemical process, we can exactly copy the operations of nature; for the artificial Seltzer waters can be made in every respect similar to those of nature.

Evidently by the early 1800s scientific knowledge and commercial activity were intersecting to promote new ways of experimenting with water's consistency, to render it more palatable and interesting to those consumers still prone to dismissing water as a bland consumable.

Although a lineage of fizzy water producers emerged, it was Genevan watchmaker and amateur scientist Jacob Schweppe

Soda fountain at Zaharako's Ice Cream Parlor, Columbus, Indiana.

who first truly commercialized the product, and whose name is still commonly associated with the drink. When he first established his Schweppes Company in Geneva in 1783, it proved unsuccessful. However, when Schweppe moved to London to market his product elsewhere, Erasmus Darwin, grandfather of the evolutionist Charles Darwin, began to talk positively to his notably wide network of prominent friends about the new, exciting beverage he had tasted. Soon afterwards fizzy drinks rapidly acquired popularity. When William IV of Britain later endorsed the product, fame was in many ways inevitable for Schweppe. The temperance movement also proved instrumental in promoting the consumption of aerated waters like those produced by Schweppe by selling them in their numerous non-alcoholic public houses.

Carbonated water was also popularly known as soda water. As inexpensive ways to bottle soda water became feasible, it became possible to transport the product across long distances. This was particularly useful in a period of imperial expansion and with the public's increasing appetite for international adventure. When George Parbury penned his handbook for travelling through India and Egypt in 1841, he wrote:

> A good supply of bottled water is an item in the traveller's wants to which the strictest attention should be paid; he must not only supply himself with a sufficient quantity, but be particularly careful that the bottles have been previously thoroughly cleaned and well cooked. Although there are places in the desert in which tolerable water is procurable, they are extremely rare; a small quantity of powdered alum should be taken for its purification – a quarter of an ounce is sufficient to clarify seven gallons. Soda water will be found a very great luxury.

The soda siphon (left), also known as the seltzer bottle or siphon seltzer bottle, is a device for dispensing carbonated or soda water. SodaStream (right), like a soda siphon, carbonates water by adding carbon dioxide from a pressurized cylinder.

The English politician Charles Wentworth Dilke once recollected that when soda water was brought to India, Indian natives believed it to be bottled British river water, envisioning Britain as a country of fizzy rivers. However, British imperialists quickly brought this superstition to an end by revealing to them that soda water was in fact technologically manufactured.

In England instilling water with gas could sometimes prove risky. How best to keep water fizzy while transporting it to shops and other vending agencies was a conundrum that deeply troubled early water producers. Fizzy water, like still water, tended to be retailed in glass bottles, although efforts to retain the fizz created recurrent problems. What became known as a Codd-neck glass bottle was the most common means of storing carbonated water until relatively recently. Designed by the British soft-drink maker Hiram Codd, the uniquely shaped bottle assisted in further popularizing fizzy water around the world.

Codd maintained the gas pressure of his fizzy water by inserting a marble in its opening. However, these bottles sometimes exploded and caused injuries as shards of glass were sent flying through restaurants and shops. Codd-neck bottles are now collectors' items that sell for thousands of pounds; they are rare because children smashed them to obtain the marbles trapped in the bottles. It was only in the 1940s that plastic bottles began to be commercially produced, and in the 1960s that production costs reduced to such an extent that bottling water in plastic on a mass scale became commercially feasible. The development of plastic bottles helped to further commercialize bottled water, although fears emerged about the introduction of BPA, or Bisphenol-A, into the drink as a result of plastic bottles. BPA is a chemical that many medical scientists link to health problems, including cancer.

Inventors and entrepreneurs also developed other novel means to enhance water's appeal as a consumable product. Fizzy water has been produced in homes since Charles Plinth first developed the soda siphon in 1813. This device allowed carbon dioxide to be added to water in controlled quantities, although having to refill the bottles at the nearest manufacturing outlet proved inconvenient for some. In 1832 the

British-born inventor John Mathews developed an apparatus that allowed drugstores and street vendors to retail cups of artificially carbonated water. Mathews's fountains produced carbon dioxide internally by fusing together the properties of sulphuric acid and calcium carbonate. The gas generated was directed to a tank of water. Some decades later, vendors in major cities around the world distributed water from these fountains. However, like Codd bottles these fountains were prone to exploding if the chemicals necessary to produce gas were inadequately mixed. Improper mixing of the acids was also feared to contaminate carbonated waters.

Although public opinion generally favoured soda water, occasional complaints surfaced, including one by Mark Twain, who asserted:

> I never enjoyed myself more in my life. I drank thirty-eight bottles of soda water. But do you know that this is not a reliable article for a steady drink? It is too gassy. When I got up in the morning I was full of gas and as tight as a balloon. I hadn't an article of clothing that I could wear except my umbrella.

Remarking on the blandness of soda water in comparison to alcohol, the nineteenth-century Romantic poet Lord Byron satirically announced, 'let us have wine and women, mirth and laughter. Sermons and soda water the day after.'

These sceptical perspectives did little to stem the rising popularity of fizzy water. When the *Evening Mail*, a New York newspaper, investigated the startling popularity of the drink in that city during 1862, its journalists estimated that around 700 New York citizens per day made a living from retailing soda water, selling between 100 and 3,500 glasses daily at the peak of summer. Soda water remained immensely popular

throughout America even as access to safe drinking water in the home increased during the twentieth century.

Despite the development of new techniques of manipulating water, a serious problem still needed to be surmounted: for some, even fizzy water tended to taste somewhat bland. Historically a number of individuals sought to overcome this obstacle by adding flavouring to both still and fizzy water to render it more palatable, as well as more marketable. Walter Hamilton recorded in 1827 that 'a pleasant saline draught is made by dissolving thirty grains of carbonate of soda or potash, and twenty grains of citric acid (acid of lemons) in two separate glasses, mixing them, and then drinking them in a state of effervescence.' Hamilton was essentially referring to an early form of lemonade, a product that consists primarily of water.

The practice of adding flavouring to water originated in medical practice. Eighteenth-century American pharmacists regularly added herbs to plain mineral waters as part of their efforts to heal and cure. The most popular additions included dandelion, sarsaparilla, birch barks and fruit extracts. The physician Philip Syng Physick of Philadelphia reportedly invented the first flavoured carbonated soft drink in 1807. The word 'tonic', as in 'tonic water', also has medicinal connotations. Chemists invented tonic water by adding quinine to carbonated water. Although quinine gave water a distinctly bitter taste, it proved useful in combating malaria, a disease still common in many countries today. Tonic water was essentially a tonic for the disease, although it tends to be used as a mixer for cocktails nowadays. The mixed drink gin and tonic originated in British colonial India, as gin was thought to improve the bitter flavour of tonic water. In fact the term 'Indian tonic water' is derived from the original purpose of the drink: to be consumed in tropical areas of South Asia and Africa where malaria was endemic.

Advertisement for Hoofland of Philadelphia's 'Celebrated German Tonic', 1860.

Departing from its medicinal roots, flavoured water, like mineral water, was quickly transformed into a commercial product. flavoured water-based beverages, made with soda fountains, became increasingly popular in America during the 1890s and could soon be found in major European cities, including London and Paris. Making use of similar techniques to those that had been applied to make water fizzy, carbonated fruit drinks also acquired considerable consumer popularity.

Perhaps the most famous form of flavoured water is made by SodaStream. Invented in 1903 by Giles Gilby, the beverages made by the device were initially marketed as a drink for the upper classes, although the product's popularity quickly filtered down to all classes. In the 1920s Soda Stream added flavour to its fizzy waters, including orange and cola. Uniquely SodaStream could be produced at home, meaning that customers no longer had to go to a bar or pharmacy to purchase a flavoured water. SodaStream have frequently marketed their products as environmentally safe, since its production involves no plastic or cans.

In the twentieth century pop (or soft water-based drinks) became ubiquitously popular in most Western countries, and was successfully marketed as a water-based product that did not suffer from being tasteless and unappealing. As one con-tributor to American magazine *The Rotarian* asserted in 1954, 'pop does quench thirst, but plain water reputedly will do the same thing. Drinking water, however, lacks that extra bit of glamour, such as spices American life at every turn.'

Squash, or cordials, also became popular as a consequence of water being increasingly recognized as a safe product in the nineteenth century. The earliest known cordials were manufac-tured in Renaissance Italy, although these tended to be alcohol-based. Modern, water-based cordials originated in 1867 when Lauchlan Ross, a descendant of a wealthy Scottish family of

ship-builders, patented a method of preserving citrus juice without the use of alcohol. He originally intended this as an aid for sailors who were prone to suffering from scurvy due to a lack of access to foodstuffs with high vitamin C content. The first factory designed to produce lime juice was established in Leith, Scotland, in 1868, and was later purchased by the Schweppes Company. Cordials proved particularly popular in the United Kingdom and Commonwealth countries. Nowadays Ribena, a concentrated drink that tastes of blackcurrants, is widely consumed in these countries. Similar to lime cordial, Ribena was developed to supply sailors fighting in the Second World War with vitamin C once orange imports were cut off by German U-boats.

The addition of ice to cold drinks also did much to reshape culinary customs globally. The question of how to store ice proved perplexing until the nineteenth century. Ancient civilizations such as Persia developed techniques of storing ice in large underground spaces in deserts. These were constructed with thick walls designed to keep the scorching desert heat out. However, it was only relatively recently that ice came to be conceived of as a dietary article that improves the experience of drinking.

Like many other water-based innovations, the modern use of ice cubes stems from medical practice. American physician John Gorrie essentially invented ice cubes in the 1840s when he built a refrigerator to cool air for yellow fever patients. Gorrie is also reputed to have invented the ice-cube tray, while his casenotes reveal that he served ice to his patients in their drinks.

Contemporaneously efforts were being made to permanently alter the ways in which ice was collected, distributed and used. Entrepreneurs have long profited from finding ways to store ice collected from rivers in winter. The u.s. businessman

Frederic Tudor's suggestions that ice could be mass-marketed, and that the general public would express an interest in buying or using it, were initially met with scepticism. Tudor perceptively recognized that adding ice to water made it more palatable and envisioned that ice could be highly marketable in the West Indies, as the stifling heat of that region made drinking water unpleasant. Although Tudor planned to ship ice from New England to the Caribbean, amused ship-owners in Boston initially refused to transport the ice he had pains-takingly collected, a predicament that forced him to purchase his own ship in 1806. But even when he arrived in Martinique, interest in purchasing ice was virtually non-existent despite Tudor's insistence that it would help to make the Caribbean heat bearable.

Despite these teething problems, using ice to cool drinks eventually became accepted as a well-established practice. In many ways Tudor kickstarted a profitable ice industry that created numerous jobs and garnered attention until well into

Loading ice, U.S. c. 1910–15.

Heurich ice plant, Washington, DC, 1919–20.

the twentieth century. His activities are important to the history of drinking water, as they profoundly altered the ways in which frozen water was commonly used both inside and outside the home, in addition to changing how water itself was drunk. In the eighteenth century only the elite classes tended to add ice to their drinks to make them more refreshing. Ice was typically harvested in winter and stored in covered wells throughout summer, although it had to be painstakingly picked with hand axes and saws, and was expensive to purchase.

The following century witnessed a rise in the popularity of the use of ice to preserve food. Certain kinds of ice, such as that procured from Wenham Lake, Massachusetts, transformed into fashionable commodities, particularly among the higher ranks of London society. Tudor founded the Wenham Lake Ice Company to cater for this increased demand for ice at fashionable and extravagant high-society dinner parties. One

successful gimmick deployed at the company's London shop involved depositing a large block of ice in the shop window and placing a newspaper on the other side of that block so that passers-by could read it through the ice. This marketing technique proved effective and garnered significant public interest, as most Londoners had never before seen a block of ice.

The early twentieth-century development of mechanical refrigeration further transformed how ice was used, although this ultimately resulted in the collapse of the nineteenth-century ice industries. The Australian journalist and politician James Harrison initially developed refrigeration, while scientist and inventor Thaddeus S. C. Lowe produced the first commercial ice machine in Dallas. Processed ice gained a deserved reputation for being healthier and safer, as it eliminated the potential of ice to be drawn from polluted, contaminated waters containing sewage or animal matter. The American manufacturer Fred Wolf pioneered modern techniques of freezing water. In 1914 he invented an unsuccessful refrigerating machine. Although relatively unremarkable, his experiments are of interest as he added a simple ice-cube tray to his design, and inspired later manufacturers to incorporate these.

From the 1920s most electric refrigerators came equipped with an ice-cube compartment. Rescuing the frozen water from the tray proved somewhat difficult until the household appliance manufacturer Guy Tinkham developed a flexible tray that allowed users to easily eject cubes. Further designs included removable ice-cube makers with release handles and eventually the plastic ice-cube trays commonly used today. Modern freezers often come equipped with automatic icemakers and dispensers built into their doors.

Evidently the nineteenth and twentieth centuries witnessed the emergence of novel ways of producing, commercializing

Photo taken for the u.s. Office for Emergency Management showing readers how to turn their freezer dial down to save electricity, February 1942.

and marketing water. With the troublesome matter of securing access to safe, clean and pure drinking water a problem of the past (at least in many Western countries), scientists and manufacturers began to investigate new ways of distributing and popularizing drinking water. The nineteenth-century public had been regularly encouraged to consume water due to the moral pressure placed upon them by temperance advocates. Over time, however, consumers became subject to the profit-making influence of drinks companies, which sought to promote bottled water for commercial rather than medicinal or therapeutic reasons. Those who were still averse to consuming and ingesting a bland, flavourless substance found themselves enticed by a range of new ways of drinking water. The introduction of fizzy waters, flavoured waters and cordials suggests that water had to be somehow manipulated to enhance its appeal, especially if the public were to willingly

purchase a natural substance that, particularly in the twentieth century, tended to be freely available at home. Ice, too, transformed into a commodity, and also served the purpose of cooling water and making it refreshing.

9
Global Access to Drinking Water

We must treat water as if it were the most precious thing
in the world, the most valuable natural resource. Be
economical with water! Don't waste it! We still have time
to do something about this problem before it is too late.

Mikhail Gorbachev

So warned Russian president Mikhail Gorbachev in response
to what he perceived to be an ominous threat of global water
shortages. The idea that parts of the planet might run out of
water seems somewhat odd given its global omnipresence.
Yet the tripling of the world's population in the twentieth century generated anxiety about the sustainability of Earth's water
resources. This fear has been intensified by high water usage
in domestic, industrial and agricultural contexts in Western
societies which placed further strain upon water supplies. The
concept of 'water stress' refers specifically to a distinct global
imbalance between water use and water resources.

Disparities in gaining access to palatable water in Western
and non-Western countries also emerged in the twentieth
century. As earlier chapters demonstrated, Western societies
historically developed an impressive array of scientific and
governmental techniques, with the intention of ensuring

Poster for the New York City Department of Water Supply, Gas and Electricity, 1941–3.

public access to safe drinking water. Yet the forms of access that evolved were not equally dispersed internationally, meaning that many countries still suffer from an inability to procure safe water supplies. Nowadays if governments advise us to cut back on our water usage or impose hose bans we express anger and bemusement at the thought of something that flows so freely from our taps potentially running out. A very different scenario exists in many countries that suffer from severe access restrictions. Paolo Lugari, founder of the Gaviotas community, an eco-village in Colombia, recently argued that 'civilization has been a permanent dialogue between human beings and water.' Lugari's statement is insightful. Levels of access to drinking water have, in many ways, played an important role in defining how different global regions have evolved and developed.

Securing safe drinking water clearly requires some degree of human agency. Water sources are often naturally polluted, contaminated and unsafe and require technical modification to render them palatable. The Earth's surface is covered with oceans, rivers and lakes. Yet around 97 per cent of the planet's

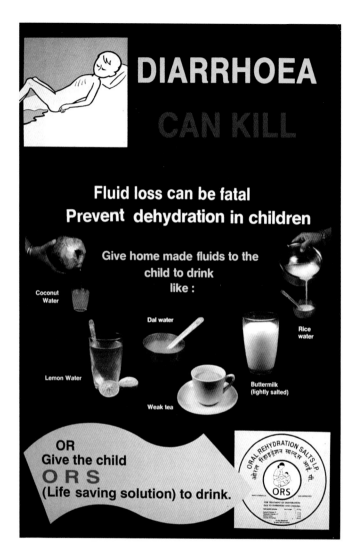

Ministry of Information and Broadcasting poster, New Delhi, 1993.

A newly installed pump on a snowy windswept plain, early 20th century.

water content is unpalatable and dangerous to consume unless technologically treated. Fresh, palatable water can be obtainable from sources such as groundwater, rainwater, surface water and springs, but these resources are relatively few and far between for many. To a certain extent, consumable water can be considered as more of a man-made product than the natural source it is often presumed to be. Yet we remain restricted in

our capacity to undertake feats such as converting seawater into drinking water, despite centuries of innovation. Efforts to do so have recurred sporadically over time. However, seawater conversion does not take place on a large scale due to the expense of desalination, the large amount of energy required and the potential damage caused to sea life.

Throughout the twentieth century, Western governments strove to ensure that safe drinking water would be available to all citizens. In 1909 the American geologist and anthropologist William John McGee asserted that 'nations which fail to conserve their water supplies have already begun to decay. The community that has an abundance of pure drinking water will rear a vigorous and stalwart race.' Hence the ability of certain societies to procure safe water has, according to some, provided a marker of civilized life and an index of socio-cultural and technological advancement. As demonstrated throughout this book, this increase in access to water in Western societies occurred during a relatively short time span. When the American writer James David Corrothers penned his autobiography in the opening decades of the twentieth century, he recalled of his youth that

> St Louis city water, in those days, looked as if it had been taken from the famously muddy Missouri river. In a single glass of it there was as much yellow mud as if a handful of mustard had been thrown into it. Housewives could not use it for washing purposes until it had been allowed to settle overnight. I felt that it must be unhealthy, and declined to become a citizen of the great river city, not stopping to consider that, if others lived there, I could.

The late twentieth-century author Edmund J. Love also later recounted an interesting anecdote about the difficulties

faced in implementing schemes designed to widen access to water in America. He recalled that

> The original plan, at the time Flushing's water system was installed, was to use the water from the river. A purification plant was built next to the pumping station, and, at the inception of service, it seemed adequate. Unfortunately, in the ten years or more between the time the first pipe was laid and the completion of the system, the city of Flint, just ten miles up stream, began to grow by leaps and bounds. The city fathers of Flint made no provision at all for growth. They simply dumped the sewage and an increasing amount of industrial waste into the river. By the time Flushing was to run its first water through the new pipes the river was hopelessly polluted . . .
>
> When the federal government finally stepped in to stop the pollution of Saginaw Bay, Flushing was awarded substantial damages and all of this money was used to drill new wells east of town. In the meantime, many years had gone by and the old water had been in the mains so long that it was deemed advisable to wait until the pipes had been cleansed for several months before anyone started drinking from the village supply. Then one day Flushing had a flushing day. Everyone was supposed to turn on his faucets for half an hour and let the new water course through the pipes. This promptly caused the whole system to break down and the village was dry for a week . . . the failure of the faucets to deliver drinking water had a lasting effect on the way that many people lived.

Despite these initial technical difficulties, domestic access to water increased rapidly during the twentieth century in Western countries where citizens typically view

availability as a basic human right. This development was accompanied by a range of advances designed to improve the quality of domestically supplied water. At the turn of the century governments and water companies sought to make 'soft water' (water with soft mineral content) more available. 'Hard water' (water with hard mineral content) typically contains more minerals. Although not technically unhealthy hard water can damage kitchens, bathrooms and toilets. In 1903 water softeners were developed which introduced sodium ions to replace the natural minerals. This replaced harmful water ions with harmless ones, and had an important impact in removing lead, mercury and other traces of heavy metals from water. Contemporaneously Western water supplies began to be sterilized by minute quantities of chlorine gas, a substance that proved particularly effective in destroying the germs that cause typhoid and dysentery. In the 1930s the Paris waterworks department reportedly used fish to test the safety of chlorine content. Before drinking water was piped to the mains, some of it was diverted to an aquarium. If the trout in the aquarium were unaffected by the chlorine content, the water was deemed safe.

The ability to transport water via vast pipes from areas with plentiful water supplies to those that were more arid also developed in the twentieth century. When it opened in 1903, the 530-km Goldfields Water Supply Scheme in Western Australia was the largest water-supply scheme of its time. During the 1950s Americans hypothesized that water could be piped from as far away as Canada to dry southern regions of America. As one contemporary magazine speculated in 1950:

> Some years from now a thirsty housewife in Los Angeles may turn on her faucet, fill her glass and take a long drink of cool water piped down to her home from the glaciers

Goldfields water-supply pipeline, western Australia.

of Canada. It will be a mighty long drink, for it will have flowed perhaps a thousand miles, along river beds and over mountain ranges, through tunnels and across deserts to quench her thirst. This is the latest proposal in an effort to bring water to the great Southwest.

Increased government regulation was a further defining feature of the twentieth century. Although some American cities implemented drinking water standards during the First World War, federal standards were applied during the 1940s. More comprehensive regulations and standards were then introduced in the 1970s. The Clean Water Act of 1972 demanded that industrial plants improve their procedures of waste disposal in freshwater sources while the Safe Drinking Water Act of 1974 specified the precise number of contaminants allowed in water samples, allowing for increased monitoring and regulation.

Despite these developments, access to safe drinking water has remained severely restricted across many non-Western societies. At worst, some scientific studies estimate that by 2030 demand for water in some developing regions will exceed supply by as much as 50 per cent. Recent estimates compiled by Water.org suggest that 884 million people lack access to safe supplies, a figure that equates to approximately one in eight people. Each year around 3.6 million people die from water-related diseases, while related complaints, including diarrhoea, claim the lives of around 1.5 million children annually.

In the 1800s Western travellers had observed makeshift strategies of transporting water that are not too dissimilar from those evident in Africa today, which is in itself suggestive of the differential development of Western and non-Western cultures of water. When the British author and traveller James Silk Buckingham visited Baghdad during the late nineteenth century,

he observed that 'the city is supplied with drinking water from the Tigris, being brought to the houses in goats skins which are conveyed on the back of animals to every man's door.' Similarly the American traveller Jerome van Crowninshield Smith stated in his *A Pilgrimage to Egypt* that 'one of the first novelties, on sailing up the Nile, is the multitudes of females transporting water, on their heads, in large, heavy earthen jars, from morning till night, to the huts and villages.'

Modern popular images of African culture are dominated by images of water carriers. Less affluent societies have failed to develop efficient means of conveying water due in part to the notoriously high costs involved in sanitizing water and undertaking plumbing works, a scenario that proves highly problematic in less affluent nations. In consequence up to one-sixth of the world's population are forced to travel a kilometre or more to collect fresh water. In the least affluent Asian and African countries carrying water occupies the time of women and children to such an extent that it prevents them from taking up educational and employment opportunities that might better their condition.

Even those countries with relatively secure access to safe drinking water are not protected from suffering periodically from shortages, should disaster strike. For instance, when a devastating tsunami struck Japan in March 2011, traces of radiation stemming from the Fukushima Daiichi Nuclear Power Station seeped into the normally safe water supplies. Localized flooding has also recently resulted in the discovery of the deadly *E. coli* bacterium in piped water within countries with sophisticated water-supply systems, such as those of Norway and the u.s.

The American forester and conservationist Bernard Frank once said that 'you could write the story of man's growth in terms of his epic concerns with water.' Frank's statement

A newly installed pump outside a one-room schoolhouse in North Dakota, early 20th century.

touches upon the extent to which water, and in particular gaining access to safe, clean water for drinking purposes, has proven to be a central concern of Western societies as they have modernized. Water was once a substance widely feared for its potentially harmful, often invisible content. Yet it has always been simultaneously venerated for its healthy, vitalizing and cleansing properties. Hence drinking water has a highly complex and intricate history as a consumable substance. In many ways the history of drinking water is one of convincing the public why they should drink water and devising ways of making it seem interesting. We are continually bombarded with advice about the healthy properties of water. One way in which its marketability has been enhanced has been to make water more appealing: to aerate it to sell it as a fizzy product; to add flavour to it in order to attract those sceptical of water's properties; and to find new ways of marketing mineral waters in an era of doubts regarding the purity and safety of the processed water found flowing from our taps. Nonetheless in many modern Western societies we continue to take for granted our access to fresh water, drinking it from taps, purchasing it bottled from shops and giving the matter little thought when we shower, bathe or swim. In fact we tend to view access to water as an individual right, not to mention a necessity for health and well-being.

Recipes

Water has traditionally been used for cooking purposes, having been recognized as safer when boiled long before the discovery of germs and the development of the idea that disease can be spread through the consumption of unboiled water. Boiling became more popular after the invention of waterproof containers (such as clay vessels), which replaced the earlier watertight baskets made from bark or reeds. Soups are known to have been made since ancient times, although their popularity rose in the eighteenth century when 'portable soup' (an early form of dehydrated food) was invented.

Steaming has played a major role in traditional oriental cooking. Archaeological evidence has demonstrated that steaming devices have been used in China for at least 3,000 years. Steamers were made from thin cypress strips in the eighth century, although these were later supplanted with bamboo. In Africa dishes such as couscous have been prepared through steaming since at least the fourteenth century. Steaming became fashionable in Western countries in the 1980s.

Pressure cooking also has a long history. In 1679 the French physicist Denis Papin invented the steam digester to reduce the cooking time of food. To achieve this, his airtight cooker used steam pressure to raise the boiling point of water. The first pressure cooker designed for home use was invented in New York City in 1938 by Alfred Vischler.

Historical Recipes

To Distil Walnut Water

Hannah Glasse, *The Art of Cookery, made Plain and Easy* (London, 1758)

Take a peck of fine green walnuts, bruise them well in a large mortar, put them in a pan, with a handful of baum bruised, put two quarts of good French brandy to them, cover them close, and let them lie three days; the next day distil them in a cold still; from this quantity draw three quarts, which you may do in a day.

To Make Buttered Water
(or What the Germans Call Egg-soop)

Hannah Glasse, *The Art of Cookery, made Plain and Easy* (London, 1758)

Take a pint of water, beat up the yolk of an egg with the water, put in a piece of butter as big as a small walnut, two or three knobs of sugar, and keep stirring it all the time it is on the fire. When it begins to boil, bruise it between the saucepan and a mug till it is smooth, and has a great froth; then it is fit to drink. This is ordered in a cold, or where egg will agree with the stomach.

Barley Water

Four Hundred Household Recipes (Bristol, 1868)

One ounce of pearl barley, half-an-ounce of white sugar, and the rind of a lemon, put into a jug. Pour upon it one quart of boiling water, and let it stand for eight or ten hours; then strain off the liquor, adding a slice of lemon if desirable. This infusion makes a most delicious and nutritious beverage, and will be grateful to persons who cannot drink the horrid decoction usually given. It is an admirable basis for lemonade, negus, or weak punch, a glass of rum being the proportion for a quart.

Raspberry Water

John S. Skinner, *The American Farmer* (1829)

Take any quantity of the ripest raspberries (or strawberries or cherries if preferred). Squeeze them through a linen cloth to extract the juice from them. Put the juice in a glass bottle uncorked, placed in the sun or on a stove until it is cleared down. Gently pour the mixture gently into another bottle without disturbing the sediment.

To half a pint of this put one quart of water, and sugar to your taste. Pour it from one vessel to the other, strain it. Put it in ice to cool.

Lemonade Water

John S. Skinner, *The American Farmer* (1829)

Dissolve one pound of loaf sugar in two quarts of water. Grate over it the yellow of five large lemons. Mix in twelve drops of essential oil of sulphur. Cut thin some slices of lemons, add, and keep the drink cool.

Apple Water

Four Hundred Household Recipes (Bristol, 1868)

Apple Water is very delicate. Cut two large apples in slices, and pour one quart of boiling water on them – or on roasted apples; strain in two or three hours and sweeten slightly.

Or:– Peel and quarter four large rennet apples, or any other firm acid apples; put them in one quart of water, with the peel of half a lemon and a handful of washed currants; let all boil for one hour, then strain and add sugar to taste. Let it remain till cold. A little wine may be added to it when about to be drunk.

Bottled Soda Water

Sarah Annie Frost, *The Godey's Lady's Book* (Philadelphia, PA, 1870)

Dissolve one ounce of the carbonate of soda in a gallon of water, put it into bottles, in the quantity of a tumbler full or half a pint to each; having the cork ready, drop into each bottle half a drachm of tartaric or citric acid in crystals, cork and wire it immediately and it will be ready for use at any time.

Modern Recipes

Fruit Water

Add a handful of fresh mint to a jug of water. Squeeze the juice of one lemon into the jug. Add a few slices of thinly cut cucumber, and serve.

Fruit Water (ii)

Add thin slices of a combination of apple, lemon, orange, pear and strawberry to a large glass pitcher. Add cold water. Refrigerate for two hours and serve over ice.

Cucumber and Rosemary Water

Cut ten cucumber slices thinly. Mix with one branch of fresh rosemary in a jug. Fill the pitcher with ice cubes and water. Leave in the fridge for at least 24 hours, and serve.

Orange and Green Water

Slice one medium-sized orange. Coat three slices of green apple with lemon juice. Mix the slices of apple and orange with the juice of half a lemon in a large pitcher. Fill the pitcher with ice cubes and water. Leave in the fridge for at least 24 hours and serve.

Herb-infused Water

Place four thin slices of lemon and twelve thin slices of cucumber in a jug of water. Add four sprigs of mint and four sprigs of rosemary. Cover and chill for up to eight hours, and serve with ice cubes.

Cucumber-melon Water

Place ten slices of cucumber and two thin slices of melon in a glass pitcher. Add water. Refrigerate for two hours, then serve with ice. Garnish with skewered melon balls.

Herb and Berry Water

Add one cup of fresh blueberries to a large pitcher of water. Add three sprigs of fresh rosemary. Refrigerate for two to four hours and serve with ice.

Orange-mint Water

Add three thin slices of orange and five mint leaves to a pitcher of water. Refrigerate for two hours. Pour over ice and garnish with mint and an orange slice.

Tonic Water

Boil 350 ml of water. Once the water is hot, add 40 g of cinchona bark. Cover and simmer for 20 minutes. Remove from the heat and strain out the cinchona bark using a fine mesh strainer. Place the liquid in a plastic container and add 220 g of granulated sugar and 6.5 g (or half a teaspoon) of citric acid. Once the sugar has fully dissolved, store the tonic syrup in the refrigerator. Dilute to make tonic water.

Rhubarb Water

Place 2 lb (4.4 kg) of rhubarb in a bowl. Pour in four cups of boiling water and cover. Let sit at room temperature overnight.

The following day, strain the liquid into a saucepan, discarding the rhubarb. Add three-quarters of a cup of sugar and the juice of half a lemon. Bring to the boil for five minutes. Cool, then strain into a bottle and cork. Refrigerate for at least two hours, and serve with ice.

Jalapeño Water

Add two whole jalapeño chillies to a pitcher of water. Refrigerate for at least an hour. Serve with ice.

Tamarind Water

Remove and discard the outer rind of eight tamarind pods. Wash the pods, then break them into small pieces. Place in saucepan and add 3 cups of water. Bring to the boil, then cover and simmer for ten minutes. Remove from the heat. Cool, cover and chill overnight.

Strain and discard the tamarind. Stir ⅓ cup of sugar into the liquid until dissolved. Pour over ice.

Mineral Water Brands

Agua Mineral Salus, Uruguay
Produced by the company Salus SA. The logo of the grounds surrounding the Uruguayan mineral spring is a puma, as local legend has it that a puma once protected the spring. Once humans arrived, the puma handed over the protection of the spring to them. Drinking from the spring is said to pass the force and vitality of the puma to human consumers.

Ambo Mineral Water, Ethiopia
The market leader in Ethiopia. The source is a thermo-mineral spring, rich in calcium, magnesium, potassium, bicarbonates and carbon dioxide. The springs are situated at the crossroads of a once major ancient trade route, although it was only in 1930 that Ambo Mineral Water began to bottle and market the drink.

Apollinaris, Germany
A naturally sparkling German mineral water, now owned by Coca-Cola. The spring was discovered by chance in 1852 and was subsequently named after St Apollinaris of Ravenna, a patron saint of wine. Apollinaris water has been referenced in famous novels by Jerome K. Jerome, Arthur Conan Doyle and James Joyce.

Arrowhead, United States
A brand of drinking water that is popular in the western United States. Arrowhead Springs first achieved fame in the nineteenth

century when David Noble Smith founded sanatorium facilities nearby to treat tuberculosis patients. The Arrowhead waters became known for their supposed curative powers and later became a popular tourist resort. The Arrowhead Springs Company was founded in 1909.

Badoit, France

A mineral water drawn from natural sources at Saint-Galmier, France. It is naturally carbonated and has been bottled for commercial uses since 1838, although it was only sold in pharmacies until 1954. Badoit became part of Evian in 1971.

Borjomi, Georgia

The Borjomi springs were discovered by the Russian army in the 1820s. The widespread popularity of Borjomi waters across nineteenth-century Russia ensured that Borjomi transformed into a popular tourist destination. Although bottling for commercial uses commenced in the 1890s, it was only when Russia took control of Georgia following the Russian Revolution of 1917 that Borjomi water was nationalized and made into a leading Soviet export.

Buxton, England

The natural mineral water of Buxton emerges from a group of springs at a constant temperature of 27.5°C (81.5°F). In the nineteenth century Buxton Spa was a popular destination. The popularity of Buxton mineral water grew considerably from 1987 when Perrier built a new bottling plant above the source and rapidly increased production.

Donat Mg, Slovenia

A natural mineral water from the springs of Rogaška Slatina, Slovenia. The spring was used in ancient times by Celtic and Roman settlers. It acquired popularity in the late seventeenth century after the Habsburg court physician Paul de Sorbait introduced the water to the court and popularized it among Vienna physicians. Donat Mg remains popular to this day.

Evian, France
Regularly portrayed as a luxury bottled water, Evian proves particularly popular among Hollywood celebrities. The Evian Company was formed in 1829. It began to be sold in its famous glass bottles in 1908.

Farris, Norway
Norway's oldest and best-selling bottled water. It acquired popularity in Norway in the late nineteenth century when doctor and health resort pioneer J. C. Holm discovered a spring rich in minerals. A spa resort was soon established around the spring and bottling commenced in 1907.

FIJI, Fiji
The company Fiji was started in 1996. The water originates from an artesian aquifer in the Yaqara Valley of Viti Levu, Fiji, although the company is headquartered in the United States.

Highland Spring, Scotland
Water has been drawn from a natural spring in the Ochil Hills, Perthshire, Scotland, since 1979, an area not actually within the Scottish Highlands. Highland Spring was the best-selling sparkling water in the UK in 2008.

Lithia, United States
The Cherokees and Muscogees have legends about their ancient ancestors who left petroglyphs in granite rock to attract future generations to the Lithia spring in Georgia, USA. A smiling turtle effigy stands over the Lithia Springs. Used until 1838 as a healing centre for the Cherokee nation, its water began to be commercially sold in the 1880s.

Malvern, England
This water has been bottled and distributed locally since the sixteenth century. It was first bottled on a large commercial scale by Schweppes, who opened a bottling plant at Holywell in Malvern Wells in 1850. Schweppes moved away from Holywell in 1890, although bottling

continued until the 1960s. In 2009 a National Lottery Heritage Grant funded the re-building of the plant, which now produces 1,200 bottles per day of spring water. The well is believed to be the oldest bottling plant in the world.

NEWater, Singapore

The brand name of reclaimed water produced by Singapore's Public Utilities Board. As a product, it consists of treated wastewater that has been purified. Although primarily used for industrial purposes, the water is potable and is used for human consumption.

Perrier, France

Drawn from a spring in the Gard Département of France. Perrier is naturally carbonated, although extra carbon dioxide is added during the bottling process to ensure that the product tastes the same as the water of the Vergèze spring.

Pluto Water, United States

A trademark for a strongly laxative natural water product popular in early twentieth-century America. Bottled in Indiana, the product was advertised with the slogan 'When Nature Won't, PLUTO Will'. It was the water's sodium and magnesium sulphate content that ensured its success as a laxative.

San Pellegrino, Italy

A naturally carbonated Italian water produced and bottled by Nestlé at San Pellegrino Terme in Lombardy. The product is generally portrayed as a luxury and expensive water.

Staatl. Fachingen, Germany

A German medicinal and mineral water discovered in 1740. The water has a high hydrogen carbonate content, meaning that physicians and their patients have regularly used it to neutralize excessive stomach acids. Staatl. Fachingen is still marketed as a product that relieves gastric discomfort.

Volvic, France
A brand of mineral water sourced from Auvergne Regional Park in France. The first of the springs was tapped in 1922. Volvic bottled water appeared on the market in 1938. Throughout the twentieth century Volvic waters became internationally known. In 1993 the Volvic Company was bought by the Danone Group. Volvic also produces a group of fruit extract drinks.

Zaječická Hořká, Czech Republic
A natural bitter water that became known for its purgative and gentle laxative effects. During the nineteenth century, bitter waters from the surrounding area were exported worldwide as the equivalent of Epsom Salt products.

Select Bibliography

Baker, Moses Nelson, *The Quest for Pure Water: A History of the Twentieth Century* (New York, 1949)

Blake, Nelson M., *Water for the Cities: A History of the Urban Water Supply Problem in the United States* (Syracuse, 1956)

Cech, Thomas V., *Principles of Water Resources: History, Development, Management and Policy* (New York, 2003)

Chapelle, Francis H., *Wellsprings: A Natural History of Bottled Water Springs* (New Brunswick, NJ, 2005)

Doria, Miguel F., 'Bottled Water versus Tap Water': Understanding Consumers' Preferences', *Journal of Water Health*, 42 (June 2006), pp. 271–6

Fagan, Brian, *Elixir: A Human History of Water* (London, 2011)

Halliday, Stephen, *Water: A Turbulent History* (Stroud, 2004)

Hamlin, Christopher, *A Science of Impurity: Water Analysis in Nineteenth Century Britain* (Bristol, 1990)

Hempel, Sandra, *The Strange Case of the Broad Street Pump* (Berkeley, CA, 2007)

Outwater, Alice, *Water: A Natural History* (New York, 1996)

Salzman, James, *Thirst: A Short History of Drinking Water* (Durham, NC, 2006)

Seaburg, Carl, and Stanley Paterson, *The Ice King: Frederic Tudor and his Circle* (Boston, MA, 2003)

United States Environmental Protection Agency, 'The History of Drinking Water Treatment', February 2000 (EPA-816-F-00-006)

Weightman, Gavin, *The Frozen Water Trade: A True Story* (New York, 2003)

Websites and Associations

Educational

AQUASTAT
www.fao.org/nr/aquastat

H2O UNIVERSITY
www.h2ouniversity.org

WATER RESOURCES OF THE UNITED STATES
www.water.usgs.gov

THE WORLD'S WATER
www.worldwater.org

WORLDBANK
www.water.worldbank.org

Conservation

ADOPT A WATERSHED
www.epa.gov/owaw/adopt

GLOBAL WATERING
www.globalwatering.com

GROUNDWATER FOUNDATION
www.groundwater.org

PROTECTING WATER
www.protectingwater.com

WATER: USE IT WISELY
www.wateruseitwisely.com

WORLD WATER COUNCIL
www.worldwatercouncil.org

Organizations

AMERICAN WATER WORKS ASSOCIATION
www.awwa.org

Global Water Supplies

CHARITY: WATER
www.charitywater.org

WATER.ORG
www.water.org

WATER FOR AFRICA
www.waterforafrica.org.uk

Bottled Water

EVIAN
www.evian.com

PERRIER

www.perrier.com

SAN PELLEGRINO

www.sanpellegrino.com

VOLVIC

www.volvic.co.uk

Acknowledgements

I would like to acknowledge my family: Pauline, Kevin, Sarah and Katie Miller, as well as Miriam Trevor. In particular, I would like to thank Kevin for his technical assistance with the photographs included in this book.

I also wish to thank Reaktion Books, as well as the series editors, for their suggestions, observations and comments, which were always helpful and insightful.

I am extremely grateful to my current colleagues at the Centre for the History of Medicine in Ireland, University of Ulster, for their ongoing support, particularly Leanne McCormick, Greta Jones and Andrew Sneddon.

This book was written while I was in receipt of an Irish Research Council postdoctoral fellowship. I would like to thank the Council and its staff members for providing the means and time necessary to write an enjoyable book.

Photo Acknowledgements

The author and the publishers wish to express their thanks to the below sources of illustrative material and/or permission to reproduce it.

Claritas: p. 62; Evian: p. 79; Freeimages: p. 86 (ilco); iStockphoto: pp. 6 (Chepko), 83 (Aleksander Kurganov); Johnnyjohnstein: p. 97 left; Laci.d: p. 77; Library of Congress, Washington, DC: pp. 22, 35, 37, 47, 59, 60, 67, 89, 95, 98, 99, 101, 104; Sean Mack: p. 110; NASA: p. 12; National Library of Medicine, Bethesda: pp. 8, 9, 14, 15, 28, 31, 34, 40, 43, 48, 49, 50, 105, 106, 113; Pastaitaken: p. 21; Adrian Pinstone: p. 65; Rama: p. 29; Shutterstock: p. 78 (lev radin); Sodastream: p. 91 right; Vmenkov: p. 17.

Index

italic numbers refer to illustrations; **bold** to recipes

Contents